GOATS OF THE WORLD

GOATS OF THE WORLD

by

Purushottam Gupta

2008

Biotech Books

Delhi - 110 035

ISBN10 81-7622-196-1
ISBN13 978-81-7622-196-2

Published by : **BIOTECH BOOKS**
1123/74, Tri Nagar,
Delhi - 110 035
Phone: 27383999
e-mail: biotechbooks@yahoo.co.in

Showroom : 4762-63/23, Ansari Road, Darya Ganj,
New Delhi - 110 002
Phone: 23245578, 23244987

Printed at : **Chawla Offset Printers**
New Delhi - 110 052

PRINTED IN INDIA

Preface

Belonging to the genus *Capra*, goats have played an important role in supplying human needs since prehistoric times, and still doso in certain areas. The goat's small size, efficiency of milk production and fertility are advantages in converting scarce plant resources into food, and as such, they are a basic component of livestock. The most important products that goats yield are milk, cheese, butter, mohair and leather. Certain goat cheeses are highly prized, as is the leather made from the skin of young goats. Because they can feed by browsing, goats are valuable in some areas for brush control. Approximately 400 million goats exist throughout the world, most of them in India, China, Turkey, Ethiopia, Iran and Brazil, and more than 200 breeds are recognised.

The present text aims to acquaint those involved in animal husbandry with the goats of the world, guiding them on the various breeds and their uses. Taking into account both their local and commercial utility, this is a handbook which enables readers, do not just recognise various goat breeds, but also there values, domestication process, management and the like. In addition, space has been devoted to an understanding of breeding techniques for increasing the quality and quantity of goat livestock numbers, and how the process is projecting future trends for goat management.

Purushottam Gupta

Contents

1

Introduction

The goat was one of the first animals to be domesticated by humans, about 9,000 years ago. There are some 200 different breeds of goats that produce a variety of products, including milk, meat, and fibre. In the last century they have also gained some popularity as pets. Female goats are referred to as does or nannies, intact males as bucks or billies; their offspring are kids. Castrated males are wethers. Goat meat from younger animals is called kid, and from older animals is sometimes called chevon, or in some areas mutton.

Goats seem to have been first domesticated roughly 10,000 years ago in the Zagros Mountains of Iran. Ancient cultures and tribes began to keep them for easy access to milk, hair, meat, and skins. Domestic goats were generally kept in herds that wandered on hills or other grazing areas, often tended by goatherds who were frequently children or adolescents, similar to the more widely known shepherd. These methods of herding are still used today. Historically, goat hide has been used for water and wine bottles in both travelling and transporting wine for sale. It has also been used to produce parchment, which was the most common material used for writing in Europe until the invention of the printing press.

Most goats naturally have 2 horns, of various shapes and sizes depending on the breed. While horns are a predominantly male feature, some breeds of goats have horned females. Polled are not uncommon and there have been incidents of polycerate goats, although this is a genetic rarity thought to be inherited. Their horns are made of living bone surrounded by keratin and other proteins and are used for defence, dominance, and territoriality. Goats are ruminants. They have a four-chambered stomach consisting of the rumen, the reticulum, the omasum, and the abomasum.

Goats have horizontal slit-shaped pupils, an adaptation which increases peripheral depth perception.. Because goats' irises are usually pale, the pupils are much more visible than in animals with horizontal pupils but very dark irises, such as sheep, cattle and most horses. Some breeds of sheep and goats appear superficially similar, but goat tails are short and point up, whereas sheep tails hang down and are usually longer.

Reproduction of Goats

Female goats (does) reach puberty at seven to ten months of age, depending on the breed and nutrition, and should be at 60 to 75% of their adult weight at breeding to prevent difficult kidding. Does will have higher lifetime production and be more profitable if they are bred to kid as yearlings. Does should kid every year thereafter until at least the age of seven or eight, if they remain healthy. Most goats are seasonal breeders, reacting to shorter days as a cue for breeding. The presence of a buck (uncastrated male goat) stimulates the reproductive cycle (estrous) and the behaviours of the does that indicate that they are in the fertile part of their cycle (in heat).

The doe's estrous cycle normally occurs from August or September until January, with October to December being the peak time for breeding. The estrous cycle is normally 18 to 22 days long. Does in heat (estrus) are at the proper stage for breeding; at this time, they will be receptive to the buck. Estrus (standing heat) lasts for 12 to 36 hours. Signs of heat include tail wagging, swollen vulva, mounting behaviour, decrease in milk yield if lactating, and a general increase in activity and bleating. Kids are born about 150 days after breeding. Planning breeding so that kids are born during the height of forage production in the spring makes efficient use of the pasture. Keeping accurate breeding records will allow you to know when kids are due and help you prepare for their arrival.

Some goat milk markets demand year-round production. Breeding season may be manipulated through the use of lights and hormone therapy. However, milk production is less for a doe that kids in the fall than when she kids in the spring. Male goats (bucks) reach puberty earlier than females and must either be separated from them by the age of four months or be castrated to prevent unwanted breedings. Buck kids can be used as herd sires at 8 to 10 months, but should not be used as heavily as mature bucks. Have your veterinarian test them for fertility and soundness before the breeding season.

The most important animal in the herd is the buck. He provides half of the genetics of the herd, and using a sound, high-quality buck can make significant improvements to the herd. Spend time and effort to locate a superior buck, one that has the traits you have identified as important. It is well worth the investment. A buck that has production records is the surest bet. At the very least, you should observe both the herd and the parents of the buck. When selecting a buck, it is important to perform a breeding soundness exam. A general physical exam can check the buck for structural soundness and abnormalities in the sex glands and organs.

The scrotal circumference should be measured, since this correlates with fertility and semen production. As a general rule, dairy bucks should measure 25 to 28 cm at 100 pounds, meat bucks should measure 26 to 29 cm at 100 pounds, and larger bucks should measure at least 34 to 36 cm. Have a semen sample taken and evaluated. A normal concentration is 2 billion sperm per cubic centimetre of semen. Of those, 70% should be motile, moving forward. The sperms' morphology should be evaluated to determine whether they are mature and whether there are abnormalities. At least 80% of the sperm should be normal. Finally, the buck's libido should be monitored. A sound buck is of no use if he will not service does. A full-grown, healthy buck should easily service up to 50 does. Bucks should not be bred to their daughters; inbreeding tends to expose genetic problems and lead to weaker stock.

Some goat producers use artificial insemination (A.I.) for breeding. This requires excellent heat detection skills and is more labour-intensive than natural service, but A.I. allows the economical use of outstanding sires. The American Dairy Goat Association (ADGA) offers a booklet about A.I., classes are offered by Langston University (Oklahoma) and sometimes by goat associations, and A.I. technicians are available in most areas of the country.

Digestive System

Goats are reputed to be willing to eat almost anything. The digestive systems of a goat allow nearly any organic substance to be broken down and used as nutrients. Contrary to this reputation, they are quite fastidious in their habits, preferring to browse on the tips of woody shrubs and trees, as well as the occasional broad leaved plant. However, it can fairly be said that goats will eat almost anything in the botanical world. Their plant diet is extremely varied and includes some species which are toxic or detrimental to cattle and sheep.

This makes them valuable for controlling noxious weeds and clearing brush and undergrowth. They will seldom eat soiled food or contaminated water unless facing starvation. This is one of the reasons why goat rearing is most often free ranging since stall-fed goat rearing involves extensive upkeep and is seldom commercially viable. Goats do not usually consume garbage, tin cans, or clothing, although they will occasionally eat items made primarily of plant material, which can include wood. They have an intensely inquisitive and intelligent nature: they will explore anything new or unfamiliar in their surroundings. They do so primarily with their prehensile upper lip and tongue. This is why they investigate items such as buttons, camera cases or clothing by nibbling at them, occasionally even eating them.

The digestive physiology of a very young kid is essentially the same as that of a monogastric animal. Milk digestion begins in the abomasum, the milk having bypassed the rumen via closure of the reticular/esophageal groove during suckling. At birth the rumen is undeveloped, but as the kid begins to consume solid feed, the rumen soon increases in size and in its capacity to absorb nutrients. Goats will consume, on average, 4.5 units of dry matter per 100 units of body-weight per day.

Uses of Goat

A goat is useful to humans both alive and dead, first as a renewable provider of milk and fibre, and then as meat and hide. Some charities provide goats to impoverished people in poor countries, because goats are easier and cheaper to manage than cattle, and have multiple uses. In addition, goats are used for driving and packing purposes. For instance, the intestine is used to make "catgut", which is still in use as a material for internal human sutures. The horn of the goat, which signifies wellbeing (Cornucopia) is also used to make spoons etc.

Goat Meat

The taste of goat meat is similar to that of lamb meat; in fact, in some parts of Asia, particularly India, the word "mutton" is used to describe both goat and lamb meat. However, some feel that it has a similar taste to veal or venison, depending on the age and condition of the goat. It can be prepared in a variety of ways including stewed, curried, baked, grilled, barbecued, minced, canned, or made into sausage. Goat jerky is also another popular variety. In India, the rice-preparation of mutton biryani uses goat meat as its primary ingredients to produce a rich taste. "Curry goat" is a traditional West Indian dish.

Nutritionally, goat meat is healthier than mutton as it is lower in fat and cholesterol, and comparable to chicken. It also has more minerals than chicken, and is lower in total and saturated fats than other meats. One reason for the leanness is that goats do not accumulate fat deposits or "marbling" in their muscles; chevon (goat meat) must ideally be cooked longer and at lower temperatures than other red meats. It is popular in the Middle East, South Asia, Africa, northeastern Brazil, the West Indies, and Belize. Chevon, as yet, is not popular in most western nations, though it is among the fastest growing sectors of the livestock industry in the US.

Other parts of the goat including organs are also equally edible. Special delicacies include the brain and liver. The head and legs of the goat may be smoked and used to prepare unique spicy dishes and soup. One of the most popular goats grown for meat is the South African Boer, introduced into the United States in the early 1990s. The New Zealand Kiko is also considered a meat breed, as is the Myotonic or "fainting goat", a breed originating in Tennessee.

Milk, Butter and Cheese

Some goats are bred for milk, which can be drunk fresh, although pasteurisation is recommended to reduce naturally occurring bacteria such as Staphylococcus aureus and Escherichia coli. If the strong-smelling buck is not separated from the does, his scent will affect the milk. Goat milk is commonly processed into cheese, goat butter, ice cream, cajeta and other products. Goat milk can successfully replace cow milk in diets of those who are allergic to cow milk. However, like cow milk, goat milk has lactose (sugar) and may cause gastrointestinal problems for individuals with lactose intolerance. It also contains a form of Casein, a Protein in all Mammal's milk.

Goat's milk is naturally homogenised, which means the cream remains suspended in the milk, instead of rising to the top, as in raw cow's milk. If the milk is kept scrupulously clean, it can be drunk raw, greatly improving the benefits. Many dairy goats, in their prime, average 6 to 8 pounds of milk daily (roughly 3 to 4 quarts) during a ten-month lactation, giving more soon after freshening and gradually dropping in production toward the end of their lactation. The milk generally averages 3.5 percent butterfat. A doe may be expected to reach her heaviest production during her third or fourth lactation. It is also said that "formula derived from goats' milk is unsuitable for babies who are lactose intolerant as it contains similar levels of lactose to cow's-milk-based infant formulae."

Goat butter is white (compared to yellow butter from cow's milk) because the goats produce milk with the yellow beta-carotene converted to a colourless form of vitamin A. Goat cheese is known as chevre in France, after the French word for "goat". Some varieties include Rocamadour and Montrachet. Feta is a well-known Greek variety that may be made with a blend of goat and sheep milk.

Fibre Production

Some goats are bred for the fibre from their coats. Most goats have softer insulating hairs nearer the skin, and longer guard hairs on the surface. The desirable fibre for the textile industry is the former, and it goes by several names. The coarse guard hairs are worthless as they cannot be spun or dyed. The proportion and texture varies between breeds, and has been a target of selective breeding for millennia. The Cashmere goat produces a fibre, cashmere wool, which is one of the best in the world. It is very fine and soft.

Most goats produce cashmere fibre to some degree, however the Cashmere goat has been specially bred to produce a much higher amount of it with fewer guard hairs. The Angora breed produces long, curling, lustrous locks of mohair. The entire body of the goat is covered with mohair and there are no guard hairs. The locks constantly grow and can be four inches or more in length.

Angora crossbreeds, such as the pygora and the nigora, have been created to produce mohair and/or cashmere wool in a smaller, easier-to-manage animal. Goats do not have to be slaughtered to harvest the wool, which is instead shorn in the case of Angora goats, or combed, in the case of Cashmere goats. However, the Angora goat usually gets shorn twice a year with an average yield of about 10 pounds while the Cashmere goat grows its fibre once a year and it takes about a week to comb out by hand, yielding only about 4 ounces. The fibre is made into products such as sweaters and doll's hair. Both cashmere and mohair are warmer per ounce than wool and are not scratchy or itchy or as allergenic as wool sometimes is. Both fibres command a higher price than wool, compensating for the fact that there is less fibre per goat than there would be wool per sheep.

In South Asia, cashmere is called "pashmina" and these goats are called pashmina goats. Since these goats actually belong to the upper Kashmir and Laddakh region, their wool came to be known as "cashmere" in the West. The pashmina shawls of Kashmir, with their intricate embroidery, are very famous.

Sustainable Production of Goats

Raising goats can be a valuable part of a sustainable farm. Integrating livestock into a farm system can increase its economic and environmental health and diversity, thereby making important contributions to the farm's sustainability. Goats often fit well into the biological and economic niches on a farm that otherwise go untapped. Goats can be incorporated into existing grazing operations with sheep and cattle, and they can also be used to control weeds and brush to help make use of a pasture's diversity. Erosion on land used for row crops declines when the land is converted to pasture. Rotating row crops and pasture every year or two offers both fertility and pest control advantages. Goats eat the forages, the goats' manure replaces some purchased fertilizers, and the life cycles of various crop and animal pests are interrupted. Like other ruminant animals, goats convert plant material that is unsuitable for human consumption into high-quality animal products.

Selection of Goats

When selecting animals for your herd you must first decide what traits are important to you and what the animals will be used for. Find a producer with the type of animals that you are interested in. You may locate producers by contacting your local Extension agent, searching classified ads in goat publications, contacting goat clubs or associations, or by attending meetings or seminars for goat producers. Once you have found a producer with goats for sale, visit the farm to observe the herd and the management. The animals will adapt more easily to your farm if their prior management and environment are similar to yours.

To develop a productive herd it is imperative that you select healthy animals. Never build your herd with animals from the sale barn. These are often animals that have been culled by another producer. There is a reason they were culled, and you do not want to bring those problems to your herd.

Listed below are some of the signs of a healthy animal.

— Shiny coat
— Lively manner
— Easy movement
— No abscesses
— Proper conditioning
— Firm, pelleted manure
— Well-shaped udder and teats

Also, ask the producer questions such as what diseases have been problems in the herd, what is the vaccination/worming protocol, and what criteria are used for selection and culling. You should also ask your veterinarian about diseases that are possible problems in your area. When selecting your animals, also observe their conformation. Drawings 1 and 2 illustrate some of the characteristics of good and bad conformation in goats.

Good and Bad Conformation

Good conformation for Dairy and Meat goats for most operations, conformation is a relatively minor concern; health and soundness are much more important. He advises checking to see that the bite is correct and that the legs and feet are sound.

Feeding Ruminants

Goats are ruminants; that is, they have a four-compartment stomach designed to digest large quantities of forages. Ruminants eat quickly and swallow their food at first without much chewing. Later, they regurgitate their food and thoroughly chew it and swallow. This regurgitated food is called the cud, and healthy ruminants will spend as much time chewing their cud as they do grazing or eating hay. This is thought to be a predator avoidance adaptation, as the ruminant can find a sheltered place to peacefully chew its cud and be less vulnerable to predator attack than while grazing.

The ruminants get their name from the rumen, which is the largest compartment of the stomach and serves as a fermentation vat. The health and productivity of the goat depends on the rumen function; microorganisms in the rumen digest fibre and carbohydrates and protein to supply the animal with nutrients. Without those microorganisms, the goat will become very sick and may die. Therefore, it is of paramount importance that the animal be fed appropriately so that the ruminal organisms stay healthy. These rumen organisms require fibre, nitrogen, and energy. Roughages have higher fibre content than grains. More mature forages contain more fibre and are less digestible. Energy is provided by good-quality roughages and by concentrates.

The rumen microorganisms have preferred pH ranges; those that digest fibre best thrive in a range of 6.0 to 6.8. Rumination increases the amount of saliva, which buffers the rumen fluid and maintains the favourable pH. However, grain decreases rumination; which means less saliva reaches the rumen, and the pH decreases. Also, in the process of digesting grain, lactic acid is produced, which can further lower the pH. When a goat eats too much grain, the rumen pH can drop below 5.5, killing the normal rumen microorganisms and resulting in a very sick animal. The rumen microorganisms are "healthiest" when goats are eating good-quality forages, such as vegetative pasture. However, it is difficult to provide good-quality forages year-round. Therefore, supplementation with concentrates may sometimes be necessary.

Raising Goats on Pasture

Contrary to the popular image of goats thriving on tin cans, goats actually require a more nutritious diet than do other ruminants. Their shorter digestive system does not retain food for as long, and thus does not digest nutrients fully. This quicker digestion allows them to eat larger quantities of food to make up for their reduced absorption of nutrients, but it is goats' unique grazing behaviour that really enables them to thrive on pasture. With their small mouths and flexible lips, grazing goats are able to select the highly nutritious parts of plants and leave parts that are less nutritious. This gives them an advantage over cattle that graze by taking large mouthfuls; within that large mouthful there might be a great quantity of poor-quality forage, including some that is dead or overlay mature.

Each goat is able to consume up to 3 to 5% of its body weight in dry matter daily. To consume that amount, however, goats must be pastured in an area with a large quantity of available vegetative forage. Goats will eat less when they are moved to poor pastures.

Listed below are some of the factors that influence intake.

— Age, size, stage, and level of production of the animal
— Animal's health
— Animal's forage preferences
— Weather
— Palatability of food
— Digestibility (fibre content)
— Maturity of forage

Goats prefer browsing but will also graze on grasses and weeds. Goats are known to stand on their hind legs to reach leaves and brush. Since goats, cattle, and sheep prefer different forages, in many pasture situations these species do not compete for the same food. Therefore, they can be managed quite successfully in a multispecies grazing system, allowing the land to be used more fully and generate more income. Land grazed by both goats and cattle returns 25%

more than land grazed only by cattle. Adding goats to a grazing system will have weed control benefits. Goats will eat such weeds as leafy spurge, multiflora rose, and brambles, decreasing the need for commercial herbicides or mowing. Meat and fiber goats are particularly useful for brush control.

When grazing goats, farmers must protect their pastures from being overgrazed. There are several reasons for this.

Overgrazing forages

- — eventually kills the plants
- — reduces the longevity of the stand and exposes more soil to erosion
- — means the animals don't get enough food
- — increases the chance of goats ingesting internal parasite larvae
- — creates bare spots, creating opportunities for undesirable weeds and erosion

The end result of overgrazing is reduced performance of both the pasture and the animals, and health problems for the animals. To prevent overgrazing, farmers should be careful to understock rather than overstock land and always remove animals from a pasture when the pasture is grazed down to about 3 to 4 inches. Browse must be managed so that it is maintained and not killed. If you want long-term production of browse, you must rotate the animals and not allow the area to become over-browsed.

Fencing is the most critical factor in raising goats on pasture. There is nothing more frustrating than having to constantly chase goats back into the pasture. Fencing will also be the greatest expense, other than the initial cost of the animals. The best permanent fencing is 4-foot woven wire with barbed wire along the top. Some graziers are also successfully using four or five strands of high-tensile electric wire. Goats may have to be trained to electric fences by placing them in a small paddock to "test" the wire. Once they have been trained to an electric fence, goats can usually be controlled with two strands of wire in a cross-fence.

Electric netting is also an option for temporary or permanent fencing in management intensive grazing systems; however, several goat producers have lost animals that tangled their horns in the netting. It is very important to keep electric fences charged at 4,500 volts or more. Regular checking and testing are necessary, and any problems must be fixed promptly, or goats will escape. Goats also need shelter. They can tolerate cold weather, but goats will get chilled by wet, cold conditions. The necessary shelter or shelters depend on the producer's operation.

A dairy operation will usually have extensive barn and pen set-ups, while a large meat goat operation may use only trees in the pasture as shelter. Buildings used for shelter may be minimal, but they should be well-ventilated and clean. Barns and sheds are not the only options for shelter. There are portable shelters, moveable shades, and even old hog huts that can be used as shelters for your animals.

Supplemental Feeding

While good quality forages are usually adequate, goats may sometimes need supplemental feeding, especially during the winter. Goats need a proper balance of energy in the form of roughage or grain, as well as protein, vitamins, minerals, and clean water. Protein and energy requirements vary, depending on the type of goat and its stage of production. There is a rule of thumb for all goats: browse and pasture in the summer, hay and grain in the winter, trace-mineralised salt at all times. When breeding begins in the fall, producer Sue Drummond feeds her angora goats not only hay, grain, and salt but also vitamins (A, D, and E) and di-calcium phosphate. Kelp, a seaweed high in minerals, is sometimes used as a supplement, though it is expensive. Alternative feeds such as roots and tubers may be fed for the energy content of the roots or the nutritious green tops. Various milling by-products are commonly fed to goats as well.

Grain is the concentrate most often fed to goats; cereal grains such as oats, corn, barley, and wheat are high in energy. Less commonplace grains such as amaranth and buckwheat are also sometimes used. Soybean meal and cottonseed meal are high-protein supplements. The choice of concentrate is determined by the composition of the forage. High-quality forages usually have adequate or even excess protein; animals eating these will need a higher-energy concentrate to utilise the protein present in the forages. Lower-quality pastures or hays will require feeding a higher-protein supplement to meet the goats' protein requirement.

Dairy goats need both high-quality forage and supplemental grain to reach their full potential, especially during peak lactation or growth. Fiber goats, on the other hand, may not do well with supplemental grain, because feeding too much protein to angora goats can make mohair fiber coarser and reduce its value, and feeding beyond maintenance requirements will not improve the fiber production of cashmere goats.

Goats can be picky eaters, and they may not immediately accept new feeds. Any feed changes should be made gradually to avoid upsetting the rumen microflora. Feeding very high levels of grain can also upset the rumen. Grain should never be more than 50% of the total diet, except for heavily-producing dairy goats. Adult meat goats should be fed a maximum of 1% of bodyweight in supplemental grain, with lactating does reaching a maximum of 1.5%.

Feeding an animal a large amount of concentrate (grain) causes acidosis: the rumen pH will drop and rumen motility will decrease. Usually the animal will go off feed, have diarrhea, and show signs of depression for a couple of days. In severe cases, acidosis can cause death. If you know an animal has consumed too much grain, you can treat it with an antacid. Call your veterinarian for help, and offer only forage and water until the animal recovers. Enterotoxemia can also occur if there is a sudden change in diet that stimulates certain rumen microbes to overpopulate and produce toxins that cause symptoms similar to acidosis. Enterotoxemia usually results in death. To prevent this disease, all animals should be vaccinated for enterotoxemia and their access to grain or lush pasture should be controlled.

Body Condition Scoring

Your goal in feeding your animals is to meet their nutritional requirements (economically) and to keep them in a productive condition. One way to monitor the animals' condition is to assign body condition scores (BCS). Body condition scoring evaluates the body fat reserves of your goats and is an easy method to evaluate the effectiveness of your feeding programme. Scores range from one to five and are determined by looking at the tail-head and loin areas. When scoring your herd, take into consideration the herd average; every herd has individuals that are too fat or too thin.

If the herd average is under or over optimal condition, usually a score of three, you need to change your feeding regimen. Body condition will vary depending on the time of year. You should try to have your animals in good body condition before winter, so they can tolerate the cold and still have adequate reserves at kidding season. The animal's stage of production also influences body condition; for example, a doe in early lactation will almost always lose condition.

Kid Management

Kids are raised for replacement stock, sold as breeding stock, or slaughtered for meat. Therefore, raising healthy, productive kids is essential to the profitability of your operation. It is crucial that kids receive colostrum soon after birth. However, in some herds Caprine-arthritis encephalitis (CAE) is a concern, and kids from those herds must be bottle-fed heat-treated colostrum instead of nursing their mothers. Kids raised naturally with their mothers usually grow better than those that are bottle-fed. However, for dairy production, it may be more economical to separate the kids from the mothers, feed kids with a milk replacer, and sell the extra goat milk. It is essential, however, that kids receive colostrum on the first day of their lives.

Males should be castrated at an early age to reduce stress on the animal. Castration with elastic bands should be done within a week of birth. There is some concern that animals may contract tetanus if they are castrated with bands. Male slaughter goats are often castrated, since the meat can have a strong flavour in intact males more than four months old. Some ethnic groups, however, want intact males. It is important for you to know your market, so you can plan for the management of your herd. Disbudding is often done in goat dairies to prevent problems with horns in the milking parlour. Kids are disbudded between three and seven days after birth, using a specially designed disbudding iron that is very hot. Equipment and instructions for use are offered by goat supply houses.

Few diseases afflict goats, and most producers find even fewer health problems when they use management intensive grazing. Practitioners of MIG see their goats at every paddock move. Observation is the best way to avoid, or at least catch early, any diseases or other problems that might occur.

When people buy goats, they should proceed cautiously.

— Check out the seller's herd.

— Ask the seller questions.

— Learn as much as they can about goats and goat diseases.

— Decide what diseases or problems they can or cannot live with, or which ones they are willing to vaccinate for or treat.

— Know what can or cannot be treated and the consequences of getting the disease in their herd.

Keeping livestock as stress-free as possible keeps their immune systems functioning properly. A healthy immune system is the best disease preventive. Conversely, periods of stress, such as weaning or transporting, may trigger disease.

Intensively managed livestock become calmer and tamer, and handling them calmly makes them easier to work with when things such as loading, vaccinating, or other tasks need to be done.

Preventive management is fundamental to maintaining health. Proper nutrition, sanitation, and ventilation, as well as timely treatment or culling of problem animals, helps keep the herd in good health and reduces health care costs. For example, the teats of milking does are usually dipped in disinfectant after milking, while the teat opening is dilated, because bacteria entering the teat can cause mastitis. Likewise, regular foot-trimming helps prevent footrot and lameness. Having a good predator control strategy will also help prevent losses in your herd.

Check with a local veterinarian to get recommendations for a vaccination and health maintenance schedule for your goat herd. Because so few medications are approved for use in goats, it is imperative to work closely with a veterinarian who can advise you on proper drug use and withdrawal times.

It is important to find a veterinarian who is compatible with you and with your management style, and who knows (or is willing to learn) about small ruminants. With time and patience, your veterinarian can become competent in the diagnosis and treatment of small ruminants.

In many areas, veterinarians recommend vaccinations for tetanus and enterotoxemia (overeating disease). Certain selenium-poor regions require the use of a selenium and vitamin injection several times a year. In other areas, additional vaccines or injections may be necessary for other diseases or deficiencies. Your veterinarian can help you set up a vaccination protocol that will protect your herd from some diseases that are problems in your area.

Managing Parasites of Goat

Parasites, especially internal ones, are the major health concern for goats. Not only are goats very susceptible to internal parasites, but the parasites are rapidly becoming resistant to all of the available anthelmintics (dewormers), and no new dewormers are being developed. Therefore, management MUST be the primary method for sustainable control of internal parasites in goats. If ample pasture is available and goats are not overstocked, a herd may have little difficulty with internal parasites. However, forcing goats to graze close to the ground and overcrowding stock will cause an increase in parasite load.

Animals on highly-stocked pastures will usually carry a heavier parasite load, due to the increased amount of faecal matter on the pasture. You can reduce parasite problems by having a low stock density and by rotating your animals to different pastures.

An understanding of how parasite infestations happen will help to avoid major problems. All parasite infestations occur when the animal ingests the infective larval stage from contaminated pasture, hay, or living quarters.

The larvae develop from eggs that were passed from an animal through its faeces. If there are no adult worms in any goats in your herd, this infestation cannot occur. Even if larvae are present in the pasture, goats are less likely than other ruminants to consume them, because goats prefer to eat at eye-level, and the larvae do not climb up grass blades to eye level. This is one of several good reasons for managing pastures to prevent grazing them too short. Try to maintain a forage height above 4 inches, at minimum.

Symptoms of a parasite problem include weight loss, rough coat, depression, and anaemia. Animals that are carrying a heavy parasite load will produce less and lag behind their herd mates. It is important to realise that heavily infected animals are "seeding" the pastures with parasite larvae, thus amplifying the problem over time by contaminating the environment.

Also, there is a great deal of variation in individual animal resistance to parasites. Culling animals with severe parasite problems will decrease the herd's problems by reducing pasture contamination and by retaining and encouraging parasite-resistant genetics in the herd.

Another way to assess the parasite load in your herd is to have a veterinarian check fecal samples for parasite eggs and recommend an appropriate dewormer, if necessary. Since very few anthelmintics are approved for goats, and since many parasites have developed resistance to anthelmintics, the help of a veterinarian is essential to administering effective anthelmintics.

For milk-producing goats, it is necessary to consider the withdrawal period that a chemical dewormer may require before the milk can be sold for consumption. Be sure to reworm three weeks after the initial treatment to kill any parasites that were ingested the day of the first worming.

Worming and then moving the goats 24 hours later will leave behind the vast majority of contaminated faeces. Pastures are considered "clean" if goats or sheep have not been grazed on them for 12 months, or if they have been hayed or rotated with row crops. In the meantime, cattle or horses may be grazed in the infested area, because they do not carry the same species of worms. Goats and sheep, however, do share the same parasites.

The main points to keep in mind about parasite control in goats are that your best defence is:

— good pasture management, including use of browse as a forage source, and

— selecting parasite-resistant animals

No dewormer will compensate for poor management, and many dewormers are no longer effective in the United States. New dewormers for goats are not being developed, so we must learn to control parasite problems through good management and selection of resistant animals.

Researchers have found that plants with high tannin levels show anthelmintic properties. The tannin in sericea l1\espedeza has been shown to suppress the egg laying ability of adult worms and inhibits the hatching of eggs that are shed. This reduces the worm load on the pasture and in the animals.

Other plants, including wormwood, may also have anthelmintic properties. Allowing the animals to graze on a variety of plant species will assist in providing better nutrition, and may also help with controlling internal parasites. Not all plant species have been evaluated to determine whether they have anthelmintic properties. The complete eradication of livestock pests is not feasible or economically necessary-some level of pests may be tolerable. Goats, like other species of livestock, may develop some immunity to worms, making a low-level infestation sometimes more advantageous than no parasites at all. Lack of immunity is very damaging to Angoras, for example.

When they are moved from arid range conditions, where there are few internal parasites, to more humid areas, where parasite population are higher, serious problems often develop. Some individual goats have a higher natural immunity than others, and those are the animals that you should select. Young goats will be most susceptible to parasites and should always be weaned to a clean pasture.

Coccidiosis, a disease resulting from infection of the intestinal tract by parasitic protozoa called coccidia, causes scours (diarrhea) in goats, particularly in kids. There are several coccidiostats on the market, but again, management is key for control. Coccidiosis occurs in damp, crowded areas. Keeping kids away from those areas prevents serious problems. Animals gain immunity to this organism by nine months of age, and clinical disease rarely occurs in adult animals.

2

Alpines

Probably one of the most striking looking of goat breeds, these handsome goats have a short glossy black coat with dazzling white face and leg markings - sandy white markings or less than jet black colouring is considered a fault. They are tall and elegant into the bargain. The hair should be sleek, smooth and short, but there is sometimes a transitory period when the winter coat is changing and the black markings are rusty, with a longer coat that is giving way to the jet black spring coat. Alpines are very good browsers and probably more content with part woodland and part pasture; if confined on a small acreage they tend to stand looking longingly over the fence.

A fault that some novices make when starting goatkeeping with this breed, lies in not supplying sufficient bulk. It can stunt the kid's growth if, from quite an early age, they are not encouraged to eat all they will of good roughage, as they have a large frame to grow. Alpines are proving their worth in commercial herds for their quality milk of high total solids that is required for today's market - something dairy product manufacturers all over the world are seeking. In New Zealand, protein tests usually average over 3.0% and milkfat tests over 4.0%. As a household milk supply, Alpines have the advantage that they milk through for two years without kidding, their milk being well known for its good flavour and high nutritional value.

French-Alpine

The French-Alpine is a breed of goat that originated in the Alps. The goats of Alpine type that were brought to the United States from France where they had been selected for much greater uniformity, size, and production than was true of the goats that were taken from Switzerland to France.

Size and production rather than color pattern have been stressed in the development of the French-Alpine. No distinct color has been established, and it may range from pure white through shades of fawn, gray, brown, black, red, bluff, piebald, or various shadings or combinations of these colors. Both sexes are generally short haired, but bucks usually have a roach of long hair along the spine. The beard of males is also quite pronounced. The ears in the Alpine should be of medium size, fine textured, and preferably erect.

The French-Alpine is a larger and more rangy goat and more variable in size than are the Swiss breeds. Mature females should stand not less than 30 inches at the withers and should weigh not less than 135 pounds. Males should stand from 34 to 40 inches at the withers and should weigh not less than 170 pounds. French-Alpine females are excellent milkers and usually have large, well-shaped udders with well-placed teats of desirable shape.

The French-alpine is also referred to as the Alpine Dairy goat and registration papers for this dairy goat use both designations and they are synonymous. These are hardy, adaptable animals that thrive in any climate while maintaining good health and excellent production. The face is straight. A roman nose, Toggenburg color and markings, or all-white is discriminated against. Alpine colors are described by using the following terms:

— Cou Blanc - (coo blanc) literally "white neck" - white front quarters and black hindquarters with black or gray markings on the head.

— Cou Clair - (coo clair) Literally "clear neck" - front quarters are tan, saffron, off-white, or shading to gray with black hindquarters.

— Cou Noir (coo nwah) literally "black neck" - Black front quarters and white hindquarters.

— Sundgau - (sundgow) black with white markings such as underbody, facial stripes, etc.

— Pied - spotted or mottled.

— Chamoisee - (shamwahzay) brown or bay - characteristic markings are black face, dorsal stripe, feet and legs and sometimes a martingale running over the withers and down to the chest. Spelling for male is chamoise.

— Two-tone Chamoisee - light front quarters with brown or grey hindquarters. This is not a cou blanc or cou clair as these terms are reserved for animals with black hindquarters.

— Broken Chamoisee - a solid chamoisee broken with another color by being banded or splashed, etc.

— Any variation in the above patterns broken with white should be described as a broken pattern such as a broken cou blanc.

British Alpine

The British Alpine was developed in Great Britian in the early

1900s. The first British Alpines arrived in Australia in 1958 and have since been graded-up using Saanen and Toggenburg does. British Alpines are tall, rangy and graceful. They are best suited to temperate climates and perform poorly in regions with high humidity. The does are good milk producers, with an average fat yield of four percent. They also exhibit a tendency to good winter milking and have an extended lactation period. British Alpines are black with white markings. Their coat is generally short, fine and glossy. However bucks often have longer hair.

3

Altai Mountain Goat

The Altai Mountain breed of wool goat was formed between 1944 and 1982 on collective farms of the Gorno-Altai Autonomous Region of the former Soviet Union; it is kept on pasture all the year round. The Altai Mountain is the result of Don goats used for improvement of the local goats. Altai Mountain goats are uniform in colour, size, conformation and wool production. They are noted for strong constitution and adaptability to the severe conditions of extensive husbandry in the highlands. They have a relatively high live weight and good meat qualities and are able to fatten rapidly during a short summer period.

Altai Mountain goats are superior to local ones in live weight (5-10 kg more) and wool yield (3-4 times more). Live weight of males is 65-70 kg and that of females is 41-44 kg. However, like local Altai goats, they mature rather slowly and grow till the age of 5-5.5 years. Wool of Altai Mountain goats has high technical qualities and is a valuable commodity for light industry. The guard hair is black and true wool dark grey. Laboratory data show that the true length of wool fibres in pedigree goats is 7.5-10 cm. The average diameter is 16-17 μm in adult goats and 15-16 μm in young animals. There is no apparent difference in wool fineness between flanks and thighs. Guard-hairs diameter is 75-90 μm. Wool content in the fleece of pedigree animals is 65-70%. Uniformity between them is not greater than 10%. Wool yields are 600-900 g for males and 450-600 g for females. Fibre length is 8-9 cm for both.

The viability of goats in conditions of extensive husba-ndry is quite satisfactory. The year-round mainte-nance on pastures does not lead to deterioration of the quality of the wool, because it grows in seasons which are favorable in terms of feeding. For Altai Mountain goats, the breaking length of wool is 9-9.5 km. For comparison, Merino wool up to 1/23 µm in diameter is considered strong enough if its breaking length is not less than 7 cm. Breeding of Altai Mountain goats is directed to enlargement of the pedigree goat population and to increase of wool production and live weight, without losing the good adaptability to the special conditions of Siberia.

Pedigree goats are also raised for sale to other farms. This work is done mainly in the collective farm 50 Years of the USSR in Kosh-Agach district and in the state farm Ediganskу in Shebalin district. In recent years this state farm has made a certain progress in raising young pedigree goats.

4

Anglo-Nubian

Anglo-Nubians was produced by crossing English goats with goats from Egypt and India. They are medium to Large animals and their eastern origins show in their Roman noses and long pendulous ears, which are set low on the head, wide and open. They are alert, sound, and well balanced with a high proud head carriage. The coat can be of any colour or combination of colours.

The breed's distinguishing characteristics include large, pendulous ears and a "Roman" nose. Due to their Middle-Eastern heritage, Anglo-Nubians can live in very hot climates and have a longer breeding season than other dairy goats. Considered a dairy or dual-purpose breed, Anglo- Nubians are known for the high butterfat content of their milk, although on average, the breed produces less volume of milk than other dairy breeds. Anglo-Nubians are large, with does weighing at least 64 kg. The average height of the breed, measured at the withers, is 81 cm for does and 94 cm for bucks. Like most dairy goats, they are normally kept hornless by disbudding within approximately two weeks of birth.

The Anglo-Nubian breed originated in England as a cross between the Old English Milch Goat and the Zariby and Nubian bucks imported from India, Russia, and Egypt. They have been exported to most countries from England and in the United States are referred to as simply Nubians. It is large in size and carries more flesh than other dairy breeds. The

Nubian breed standard specifies large size, markings can be any color, the ears are long, pendulous, and the nose is Roman. The Nubian temperament is sociable, outgoing, and vocal. Because of its elongated ears and sleek body, the Nubian is occasionally nicknamed the "Lop-Eared Goat" or "Greyhound Goat".

According to the American Dairy Goat Association the Nubian goat should be: a relatively large, proud, and graceful dairy goat of mixed Asian, African, and European origin, known for high quality, high butterfat, milk production.

The head is the distinctive breed characteristic, with the facial profile between the eyes and the muzzle being strongly convex (Roman nose). The ears are long, wide and pendulous. They lie close to the head at the temple and flare slightly out and well forward at the rounded tip, forming a "bell" shape. The ears are not thick, with the cartilage well defined. The hair is short, fine and glossy. Any color or colors, solid or patterned, is acceptable.

The Nubian's size makes it a very useful dual purpose animal. The Nubian breed leads the way for the dairy breeds in butterfat production: it produces on average, 5% or more butterfat content. This is surpassed only by the Nigerian Dwarf , Pygmy goat and Boer goat breeds, which are less likely to be used for large scale milk production, as for a dairy or cheese making. Because of the higher fat content, the Nubian milk has more flavor than lower fat milks. Milk production is lower than other dairy breeds on average. The Anglo Nubian is an all-purpose goat, useful for meat, milk and hide production. It is not a heavy milk producer but has a high average butter fat content. The Anglo Nubian breeding season is much longer than that of the Swiss breeds so it is possible to produce milk year round.

As it is the best suited of the dairy goat breeds to hot conditions, the Anglo Nubian has been used in grading-up programs in many tropical countries to increase the milk and meat production of local breeds. The Anglo-Nubian is a relatively large, proud, and graceful dairy goat. The Anglo-Nubian goat is named for Nubia, in northeastern Africa. The originally goats imported from Africa, Arabia and India were long-legged, hardy goats that had some characteristics desired by goat breeders in England. English breeders crossed these imported bucks on the common short-haired does of England prior to 1895 to develop the Anglo-Nubian goat. In the United States the breed is usually spoken of as the Nubian.

The Anglo-Nubian is regarded as an "aristocratic" appearing goat and has very long, pendulous ears that hang close to the head. The Anglo-Nubian carries a decidedly Roman nose and is always short-haired. Any solid or parti-colored coat is permitted in the Anglo-Nubian, but black, red or tan are the most common colors, any of which may be carried on combination with white. Usually there is shorter hair on the Anglo-Nubian males, particularly along the back and on the thigh, than is commonly found on the Swiss breeds.

The udder of the Anglo-Nubian is capacious but is sometimes more pendulous than that of the Swiss breeds. A mature doe should stand at least 30 inches at the withers and weigh 135 pounds or over, while the males should stand at least 35 inches at the withers and weigh at least 175 pounds. The Anglo-Nubian usually gives less milk than the Swiss breeds, but produces a milk of higher butterfat content. The head is the distinctive breed characteristic, with the facial profile between the eyes and the muzzle being strongly convex. The ears are long, wide and pendulous. They lie close to the head at the temple and flare slightly out and well forward at the rounded tip, forming a "bell" shape. The ears are not thick, with the cartilage well defined.

5

Angora

The Angora goat is a goat from the Angora region in Asia Minor, near present-day Ankara, Turkey. It is first mentioned in the time of Moses, roughly 1500 B.C. The first Angora goats were brought to Europe by Charles V, Holy Roman Emperor, about 1554, but, like later imports, were not very successful. The fleece taken from an Angora goat is called mohair. A single goat produces between five and eight kilograms of hair per year. Goat shearing is done twice a year, in contrast to that of sheep which are shorn only once.

Angora goats are more susceptible to external parasites (ectoparasites) than similar animals, because of the density of their fiber. They are not prolific breeders, nor are they considered very hardy, being particularly delicate during the first few days of life. Further, Angoras have high nutritional requirements due to their rapid hair growth. A poor quality diet will curtail mohair development. For a long period of time, Angora goats were bred for their white coat. In 1998 the Colored Angora Goat Breeders Association was set up to promote breeding of colored Angoras. Now Angora goats produce white, black, red, and brownish fiber.

The Angora dates back prior to early biblical history. Mention is made of the use of mohair at the time of Moses, which would fix the record of the Angora some time between 1571 and 1451 B.C., according to the Angora Goat Mohair Industry publication from USDA. Mohair became a valuable product in commerce early in the nineteenth century. In order to increase the supply of mohair available for export to the European countries, the Turks crossed the Angora goat with common stock to increase the poundage of salable hair. Probably there was no effort to keep the original Angora separate, and the general increase in size and vigor of the goats in the Angora area was, no doubt, partially the result of this infusion of other blood.

Angora stock was distributed to different countries, and a pair of Angoras was imported to Europe by Charles V about 1554. In 1765 an importation was made by the Spanish government and twenty years later a considerable number were imported into France. None of these importations were successful in establishing mohair production. On the other hand, Angoras were taken to South Africa in 1838, and from this importation and later importations mohair production was established in that country. The Union of South Africa is one of the three leading mohair-producing sections in the world and is exceeded in production only by the United States and Turkey.

Mohair Production

The most valuable characteristic of the Angora as compared to other goats is the value of the mohair that is clipped. The average goat in the U.S. shears approximately 5.3 pounds of mohair per shearing and are usually sheared twice a year.

They produce a fiber with a staple length of between 12 and 15cm. The mohair is very similar to wool in chemical composition but differs from wool in that it is has a much smoother surface and very thin, smooth scale. Consequently, mohair lacks the felting properties of wool.

Mohair is very similar to coarse wool in the size of fiber. It is a strong fiber that is elastic, has considerable luster, and takes dye very well. Mohair has been considered very valuable as an upholstering material for the making of plushes and other covering materials where strength, beauty, and durability are desired. The market valuation of mohair fluctuates more than does that of wool, but, in general, satisfactory prices are obtained for the clip. During depressed times, the market has favored fine hair and because fine hair is normally shorn from young goats, selection for fertility has also become increasingly important.

Head and Body Characteristics

The Angora is very picturesque animal in which both sexes are horned. The bucks usually have a pronounced spiral to the horn, which comes back and away from the head; the horns of mature bucks sometimes reach two or more feet in length. In contrast, the horn of the female is comparatively short, much smaller, and has only a very slight tendency to spiral. The horn of the female seldom exceeds nine or ten inches. The ears are heavy and drooping. The Angora goat is a small animal as compared to sheep, common goats, or milk goats.

There is considerable variation in the size of goats, but mature bucks will usually fall in a weight range of from 180 to 225 pounds but do not reach their maximum weight until after five years of age. Does will fall in a weight range of from 70 to 110 pounds when mature. The form of the goat should be similar to that of sheep, but the mutton characteristics are much less developed. The back should be straight, the rib well sprung, and the body deep and uniform in depth. The chest should be wide, the legs straight, and the rear quarters should show ample development.

Ringlet Type Goat

Modern Angoras are often classified according to the type of ringlet or lock hair in which the hair grows. Ringlet type goats are often referred to as the C Type, while B Type is used to designate those with a flat mohair lock. In the case of the ringlet type goat, the mohair is carried in tight ringlets throughout almost its entire length and represents the finest mohair produced. The flat lock, in contrast, is usually wavy and more bulky in appearance.

The Angora has a grazing habit which has made it very adaptable to certain agriculture sections. Goats are great consumers of "browse" and have a tendency to eat as high as they can reach by standing on their hind legs. This adapts goats for grazing areas on which sheep would not do well. Angoras have been able to make economic returns on land that is unsuitable for usual agriculture pursuits.

The Angora goat is not as prolific as other goats and twins are not the usual birth. Goats in large range bands will usually kid from 60 to 70 percent, but in well-managed small bands of purebred goats the rate of reproduction may be slightly over 100 percent. The Angora is among the most delicate of our domesticated animals. They are more susceptible to damage from internal parasites than are sheep. They are extremely delicate at birth, and the young need some protection during their first few days if the weather is cold or damp. Although the mature goat is fairly hardy animal when in full hair, it cannot withstand cold wet rains immediately after shearing. Storms cause excessive losses in Angora flocks at kidding time or at shearing time.

For many years, goat carcasses have been sold as low-grade sheep carcasses. Angora goat breeders, however, have sponsored an endeavor to have the Angora goat carcass referred to as "chevon' and sold as such on the market. It is maintained that the flesh is quite acceptable, particularly from young goats in good condition, and some people refer to it as a delicacy. The carcasses are characteristically thinner fleshed and the dressing percentages correspondingly lower than those of sheep that have been fed in a similar manner.

6

Appenzell

The Appenzell Goat is a pure white goat with mid-length hair. It has a powerful, harmoniously built body. The limbs are powerful and well positioned. The body is somewhat broader than that of the Saanen and therefore somewhat smaller. This goat is bred hornless. The height of the whithers is said to be 75 to 85 cm with bucks, and 70 to 80 cm with the nannies. The male animals are on average 65 kg in weight, and the female approx. 45 kg.

The Appenzell Goat is found above all in both cantons Appenzell and St. Gallen (Toggenburg). In canton Zurich only the Zurich Goat is bred (today only very little!). It is a cross between the Appenzell and the Saanen. In milk production the Appenzell Goat is on par with the other breeds: during lactation between 700 and 800 kg, fat 2.9%, protein 2,7%, lactation duration 270 days. As is generally known, milk production is heavily influenced by care. The number of herd book animals is around 800; the stock is slowly increasing again. This breed was very heavily affected by the CAE-Virus, so that extensive restoration had to be done. Today the breed is essentially CAE-free. The breed is officially recognized. The breeding books are kept by breeding associations and the herd book by the small animal center.

7

Arapawa Island Goat

The Arapawa goats are among the few survivors of the Old English milch (milk) goat which is now extinct to their native England. They may have been introduced as early as the late 1830s by the first European settlers, who established a shore whaling station on the island. Goats were recorded at the station in 1839. Left on Arapawa island to breed as a future source of food and milk by explorers, the goats came under seige in the 1970's. The goats have been recognized by the Rare· Breeds Conservation Society of New Zealand. A number of Arapawa goats have been removed from the island over recent years and are now being bred by a few enthusiasts in various places throughout New Zealand. They are also being maintained in a reserve on Arapawa Island itself. In 1993 they were exported to USA and in 2004 to Great Britain. The goats are colorful with distinctive markings and brown and black patchwork.

Without doubt the most interesting group of feral goats remaining in New Zealand is that found isolated on Arapawa Island in the Marlborough Sounds. They are a relatively small breed and come in a variety of colours - patterns of white, fawn, brown and black being common - and they usually have distinctively patterned faces. The males have widely sweeping horns, the females shorter backward-pointing horns.

It is widely believed that these goats are a surviving remnant of the Old English breed, possibly descendants of a pair released by James Cook in 1773. The colour and markings of the animals have been cited as supporting this idea. However, Cook actually recorded that the goats he released in East Bay on Arapawa Island - and which he had taken aboard at the Cape Verde Islands on the voyage out - were killed before he left the country. The fate of a second pair of goats he left on the nearby mainland in 1777 is not known - these had been collected at the Cape of Good Hope, specifically for release in New Zealand. That, however, does not obviate the possibility that the Arapawa goats are descendants of the Old English breed.

DNA analyses undertaken in Spain in 2007 indicated that the Arapawa goats were a genetically important breed in their own right and only distantly related to a number of other breeds investigated at the time. Unfortunately, no DNA material of the Old English goat was available for comparison. The goats of Arapawa Island have never numbered more than a few hundred, and were always subject to intermittent hunting. In the 1970s they came under the threat of eradication when it was thought that they were seriously damaging the island's native forest. As no acceptable proof could be produced of their antiquity or rarity - or of any immediate commercial potential - a programme of severe culling was instigated.

8

Australian Feral Goat

The feral or wild goat has been present in Australia in quite large numbers since the turn of the century. Feral goat populations started with the abandonment of goat flocks in unprofitable or drought times and through the gradual build-up of escapees from shepherded goat flocks. Over the years feral herds have developed and multiplied in areas which have offered some natural defense against predators. These goats have effectively adjusted to the environment to the degree that there is no longer any strong resemblance to the original breeds.

Research indicates that these animals possess strong fiber and milk producing qualities. When domesticated and carefully selected, they provide an excellent base for the development of cashmere, mohair, leather and meat production programs. This domesticated animal is now being referred to as the Australian goat. An increasing number of producers are running them in conjunction with sheep, as a means of controlling weed and scrub growth. These enterprises can now provide significant numbers of goats that are even in age, size and weight. Does from these flocks are suitable for breeding Cashmeres, cross-breeding with Angoras for fiber production and as recipients in fertilized ovum transplant programs.

Entrepreneurs are also utilizing the Australian goat for the overseas export of meat and skin. The feral goat has established populations in a variety of habitats across Australia. It competes with native fauna and causes land degradation, threatening plant and animal species and ecological communities. The feral goat can be an agricultural pest, but also has commercial value and is harvested for its meat. To protect the environment, feral goat control programs are best undertaken in areas of high conservation value.

Goats are considered one of Australia's worst feral animals. They remove grass from hills, therefore, causing Soil Erosion and Land Degradation. They compete with rock wallabies for food; they kill trees and bushes by eating leaves, and stripping the tree's protective bark just to name a few things these monstrous animals have done, and are still doing to the Australian environment.

Feral goats are herbivores. That means that they only eat plants. Goats, like Camels, will eat any plants and just about any part of the plant too, but feral goats tend to go for the most nourishing plants, therefore leaving the plants that other animals won't eat. An example of this is the fact that goats and rabbits have eaten the Phillip Island Hibiscus to virtual extinction. Only 2 clumps of this plant now remain. This type of Hibiscus was found only on Phillip Island, which is just off the coast near Melbourne, Victoria. In the future, cases like this could happen as a result of overgrazing. Feral goats have a massive contribution to overgrazing.

Goats are normally found in areas that also inhabit other grazing animals, including wallabies, kangaroos, and domesticated livestock. The rare Yellow-footed Rock-wallaby was nearly extinct due to the massive overlap of the diet and habitat with the feral goats. Since feral goat numbers have been heavily culled in the areas that the Yellow-footed Rock-wallaby inhabits, these marsupials have made a comeback. If the event that exotic disease would reach the Australian shores, feral goats would be prime carriers and act as disease reservoirs for many diseases, for example bovine foot-and-mouth disease, which devastated livestock in Great Britain and Europe around the turn of the last century costing billions of dollars; rabies, which can effect humans and most animals; blue tongue; rinderpest; and scrapie.

9

Bagot

This breed was formerly feral at Blithfield Hall, Staffordshire, England but has become scattered since 1957. The breed is nearly extinct. The Bagot goat is one of the oldest breeds of goats in Britain today. It goes back in history to around the 1380s. This is when it was introduced to Blithfield Hall in Staffordshire. The actual date for this is well documented. It is likely that the Bagot may be descended from the Schwarzhal goat from the Rhone valley. There are many stories as to how they got to Blithfield Hall, the most likely one which cannot be proved , but the most factual being, some goats were brought back to England by the returning crusaders.

Some were it seems, given to the then John Bagot who was at Blithfield, by King Richard II. This was in recognition of a good days hunting by the King at Blithfield, in Babots Park. It is known that these goats have been running semi-wild at Blithfield for well over 600 years. As regards the history story and the crusades. It must be pointed out that the crusades finished around 1100. The goats, so the story goes, went to Blithfield around 1380. This would indicate that the goats probably went first to one of the Royal Parks. The damage they would have done there, no doubt would have been considerable and the King would have been glad to get rid of them to another place where they could be enjoyed for hunting.

The head of a goat has appeared on the coat of arms of the Bagot family since the year 1380 and the close association of the family with goats is believed to stem from the presentation of a herd of goats to Sir John Bagot by Richard II. In the Bagot family church at Blithfield, the crest with the head of a goat is carved on tombstones and there is a stone frieze of goat heads at Goat Lodge near Blithfield Park. Available documentary evidence of the continuous existence of a herd of goats in Bagot's Park, which lies close to Abbots Bromley in Staffordshire, is sparse but the colour of the goats which is similar to that of the Schwarzhal goats in Switzerland lends credence to the theory that the Bagot goats are possibly descendants of animals brought back from the mainland of Europe, perhaps by the Crusaders. This makes the Bagot goat one of the oldest breeds of goats in Britain today.

The goats arrival at Blithfield around 1380 would indicate that they probably went first to one of the Royal Parks. The damage they would have done there, no doubt would have been considerable and the King would have been glad to get rid of them to another place where they could be enjoyed for hunting. Bagot numbers have fluctuated for a variety of reasons but the breed is now secure as herds exist in many locations, and numbers are increasing, even though it is not a productive breed, except possibly for meat. Purchasers of

Bagot stock should ensure that it is pure-bred. At one time a grading up programme was operated to increase numbers and decrease inbreeding. Bagot males were used on any female goat and progeny were backcrossed. With hindsight this programme produced unfortunate results.

The Bagot Goat is one of the oldest known registered breeds of goat in Britain today. It can be traced back in history to around 1387 when given, as believed, to Sir John Bagot who lived at Blithfield Hall in Staffordshire by King Richard II. This, we are told, was in recognition of good hunting by the King in Bagots Park. When John Bagot was made Sir John, he apparently incorporated the goats into his own crest. Also his helmet with its goats head top can still be seen to this day hanging in Blithfield church. Ever since, the Bagot family has used the goat as their mascot, incorporating it into their Coat of Arms. There is also an old legend that states that when the Bagot goats die out, so will the Bagot family die out. That may be why the herd has survived so long.

It is well known that the Bagot Goats have been at Blithfield running semi-wild for well over 600 years. The goats roamed the parkland at Blithfield and were just hunted like deer. For centuries the goats lived in Bagots Wood and Park, which was in fact some miles across the estate away from the Hall. To give you some idea of the size of the Blithfield Estate in those times, it consisted of about 5,000 acres, split up into about 2,000 acres of woodland and 1,000 acres of parkland.

The rest was mostly farmland of some sort or another. It is also known that these Bagot goats were not classed as truly wild like the deer. The reason for this is that when any of these goats got out of the fenced area and did damage to local farm crops, then the Hall had to pay compensation for the damage they had done. This system had gone on since early times. Any damage done by the deer was not the same, and nothing would be paid out as they were classified as wild. No doubt, in times long ago the goats were helped to get out on occasions, in order to get more compensation paid.

The actual numbers of goats on the Blithfield estate have varied through the ages. We do know from documentation that in 1710, for instance, that the Hall had to reduce the number of Bagot goats on the estate in the interests of agriculture. Again in 1938, we are told, when there was over one hundred in the herd, they were again reduced. We also know that in the Second World War, there was an extermination order placed by the War Agriculture Executive on the complete herd, again because of the damage they were doing to crops along with the deer. The goat numbers at that time were above eighty. The then Lord Bagot fought this order and got it altered, but had to agree not to let the herd number exceed sixty. This level was kept at sixty by culling until 1947. After this the herd was allowed to increase yet again. We know that in 1954 the number was up to around eighty. In 1954 some Blithfield stock was sent to other large country Parks.

In 1954 some male Bagot stock was released on mountains in Wales. This may be in part why some of the wild goats that are there today, have BAGOT type markings. Around this time, in 1957, the Blithfield goats' fate was about to be sealed once and for all. The (5th) Lord Bagot sold the Blithfield Estate and the goats had to go. Some were sent to Halls and Zoos and some, we are told, went to private individuals. Many just went for dog meat no doubt. However, many in the United Kingdom are now scattered in small herds, which will ensure that they survive. Nancy Lady Bagot bought twenty of the surviving goats in 1962. These went back to the Hall at Blithfield under the care of Nancy Lady Bagot. The Hall was open to the public at that time.

The Bagot goat is basically a feral animal and is therefore very wary of humans. Although of course it can be tamed, they still keep their nervous character. They are quite hardy and well able to defend their kids from any predator. In the early 1950's the Bagot goats had their territory drastically reduced. Some of the Blithfield estate southeast of the Hall was taken over by the Staffordshire Water Authority to build a massive reservoir. After that, the parkland on the estate was sold and therefore the herd had to be dispersed, with only a few going to the Hall. From then onwards the goat numbers at Blithfield slowly declined. Nancy Lady Bagot then decided to send her remaining goats to the Rare Breed Survival Trust to ensure their survival.

The Bagot Study Group was formed in 1989 by some of the breeders to try and help the breed. This group later transferred to a full Breed Society, and now it is true to say that Bagot goat numbers have dramatically increased, and the overall quality has much improved. The Bagot goat is impressive to look at and certainly has the longest recorded history of any British goat, but has no real commercial value at all. The breed standard is slowly being improved and the Rare Breed Survival Trust idea of grading has now gone out of favour. Bagot goats are mostly black at the front and white at the back with long shaggy coats. The colour change should take place around the front shoulder line. Both male and female are horned, the males have larger horns, which sweep back and turn outwards. The overall impression should convey a medium sized, strong and well-proportioned animal. Supernumerary teats are banned in males and they are to be discouraged in females.

The Bagot goat society was formed by its breeders solely to insure that the Bagot breed of goats survive and to help promote and encourage people who are interested in this rare breed of goats. Nancy Lady Bagot, owner of the Blithfield Estate, is President of the Goat Society and Peter Evans is Secretary. The Bagot goat is now documented far better than ever by the Society. The organization is really just a collection of breeders trying to help the breed to survive. The group membership is open to all those who have an interest in this animal with most Bagot breeders being members. The group sends out a breed newsletter, keeping members in touch with each other, and general publicity takes the form of shows, talks, workshops, booklets, and videos. The Bagot Breed is now totally controlled by the Bagot Breeders, including all breed registrations.

10

Barbari

The Barbari is a meat type breed that is found in Mathura District of Uttarpradesh, as well as Gujrat, Jhelum and Sargodha districts in Pubjab Province. They are a small size and their color is white creamy to golden.

Spotted animals also come across. Their meat conformation is considered good. Triple kidding and early maturity are common features of these goats.

The breed, quite important for milk and meat, is a dwarf breed highly suited for rearing under restrained and stall-feeding conditions.

In addition to being a good milkier, it is highly prolific. Considering the number of animals reported, there is serious need for undertaking conservation measures.

It was observed in personal surveys that most males are castrated early in life and fattened for slaughter at religious festivals, and that a sizeable number of non-descript goats are kept for milk production by owners who maintain Barbaris, as all the milk of the latter is allowed to be suckled by their kids to ensure good growth.

Climate

Average monthly temperature (°C)	Average	Range
Minimum	18.8	7.5-28.8
Maximum	32.3	22.2-41.7

Average monthly relative humidity (%)		
Morning	62	33.3-83
Evening	43.6	20-74
Annual rainfall (cm)		77.0

Breed Characteristics

	Adult males	Adult females
Body weight	37.85	22.56
Body length	70.45	58.68
Height at withers	70.67	56.18
Chest girth	75.53	64.31

Small animals, with compact body. The orbital bone is quite prominent, so that eyes appear bulging. There is wide variation in coat color, but white with small light brown patches ,is the most typical. Ears are short, tubular, almost double with the slit opening in front, erect, directed upward and outward.

Both sexes have twisted horns, medium in length and directed upward and backward; horn length: 11.17 cm. Bucks have a large thick beard.

The average flock contains 9.25 individuals, of which 0.6 adult males, 5.3 adult females and 3.4 young. Mostly maintained on browsing and grazing stubble of cultivated crops and tree leaves. Cultivated fodders and grains are commonly offered. The animals are generally housed in thatched sheds inside the house. Most males are castrated at ages between 7 to 30 days by the open method, with a blade, and are fattened on milk for slaughter on festive occasions.

Reproduction

	Days
Age at first kidding	648
Kidding interval	348
Service period	70
Kidding percentage (%)	70.2
Litter size: (%)	
Singles	49.64
Twins	49.32
Triplets	1.04
Quadruplets	

Mortality %

Pre-weaning	0.87
Adults	0.19

Performance

	Milk
Lactation yield (Kg)	107.120
Lactation length (days)	150.13
	Meat (kg.)
At birth	1.739
At weaning	6.661
6 months	7.800
9 months	12.566
12 months	14.517

11

Beetal

The Beetal is a breed used for meat and milk production. Found in Punjab, Pakistan and India, the Beetal is usually red, black or pied with pendulous ears. The males have long twisting horns. The breed is similar to the Jamnapari but smaller.

Throughout the States of Punjab and Haryana. True-bred animals are however found in the districts of Gurdaspur, Amritsar and Ferozepur in Punjab. The total goat population in the Beetal distribution area, according to the 1972 livestock census, was 0.159 m, of which 0.014 m adult males and 0.084 m adult females.

Climate

Average monthly temperature (°C)	Average	Range
minimum	17.3	5.8-27.1
maximum	31.9	20.2-41.2
Average monthly relative humidity (%)		
morning	66	37-83
evening	44	22-66
Annual rainfall (cm)		70.4

The goat population of Punjab has shown a marked decline. Surveys carried out by the Division of Dairy Cattle Genetics of the National Dairy Research Institute (NDRI) showed that both the number of flocks and the number of goats per flock have decreased, primarily owing to an ;increase in ;irrigated cultivated areas and the shortage of natural vegetation available for browsing. Beetal is a good dairy breed, second to Jamnapari in size but is superior to it in that it is more prolific and more easily adaptable to different agro-ecological conditions and to stall-feeding. Jamnapari is more sensitive.

Breed Characteristics

	Adult males	Adult females
Body weight	59.07	34.97
Body length	85.	70.42
Height at withers	91.60	77.13
Chest girth	86.0	73.7

Large animals. Variable coat color, predominantly black or brown with white spots of differing sizes. In a survey conducted by NDRI in the home tract of the breed, 92.6% animals were black and 7.4% brown.

The coat is short and lustrous. The face line is convex, with typical Roman nose but not as prominent as in Jamnapari. Ears are long and flat, curled and drooping ear length: 24.8 t 0.65 cm. Both sexes have thick, medium-sized horns, carried horizontally with a slight twist directed backward and upward; horn length: 11.95 t 0.76 cm . Tail is small and thin.

The udder is large and well developed, with large conical teats.

The average flock contains 21.06 + 1.92 individuals, of which 1.5 adult males, 11.7 adult females and 7.8 young.

Reproduction

	Days
Age at first kidding	761
Kidding interval	368
Service period	160
Kidding percentage	176
Litter size: (%)	
Singles	40.66
Twins	52.6
Triplets	6.52
Quadruplets	0.22

Mortality %

Pre-weaning	25.4
Adults	13.2

Performance

	Milk
Lactation yield	177.38
Lactation length	187.0
	Meat Kg
At birth	2.80
At weaning	9.26
6 months	12.18
9 months	15.42
12 months	21.83

BREEDING

Pure breeding. Breeding males are generally selected on the basis of their dam's milk yield. The services of a buck of a neighboring owner are commonly utilized when a flock does not possess a buck.

12

Begal

Individuals of this breed are sometimes distinguished by colour as Black, Brown, Gray and White Bengal. Personal surveys suggest that the breed could be separated into two types, one found in the hot humid plains of West Bengal, Bihar and Orissa, and the other in the sub-temperate and humid uplands and plains of Assam and other northeastern States. The latter were earlier known as Assam Plain and Assam Hill goats, depending on their location, but are not substantially different from the former.

Distributed throughout all eastern and northeastern India, from Bihar through northern Orissa to all West Bengal, Assam, Manipur, Tripura, Arunachal Pradesh and Meghalaya.

The total Bengal-type goat population in the States of West Bengal, Bihar and Orissa, according to the 1972 census, was 14.164 m, of which 2.198 m adult males and 6.257 m adult females. In Assam and other northeastern States, the goat population was 1.501 m, of which 0.304 m adult males and 0.543 m adult females.

CLIMATE

Average monthly temperature (°C)	Average	Range
minimum	24.4	12.4-26.3
maximum	28.7	26.3-37.5
Average monthly relative humidity (%)		
morning	73.8	61.8-84.8
evening	62.4	37.4-82.7
Annual rainfall (cm)	147.5	

BREED CHARACTERISTICS

Size

- West Bengal, Bihar and Orissa	Adult males	Adult females
Body weight (kg)	32.37 ± 2.74 (4)	20.38 ± 0.16 (554)
Body length (cm)	63.2 ± 2.14 (4)	51.2 ± 0.16 (554)
Height at withers (cm)	58.3 ± 3.16 (4)	55.4 ± 0.18 (554)
Chest girth (cm)	72.0 ± 1.35 (4)	63.2 ± 0.16 (554)
- Assam and other north-eastern States		
Body weight (kg)	15.38 ± 0.58 (25)	14.27 ± 0.20 (137)
Body length (cm)	53.8 ± 0.72 (25)	54.8 ± 0.26 (137)
Height at withers (cm)	49.5 ± 0.64 (25)	49.3 ± 0.24 (137)
Chest girth (cm)	58.7 ± 0.81 (25)	59.1 ± 0.23 (137)

Conformation

The predominant coat colour is black; brown, grey and white are also found, the former two sometimes with black markings along the back and on the belly and extremities. In a survey of 230 goats in Assam, 96 were completely black, 51 completely white, 11 brown, 45 white with black markings, and 17 brown with black markings. In another survey in Bihar, the proportion of grey to brown was 35:65. The hair coat is short and lustrous. The nose line is slightly depressed. Both sexes have small to medium horns, directed upward and sometimes backward; average horn length: 6.4 cm (range: 5.8 to 11.5 cm). Beard is observed in both sexes. The ears are short, flat and carried horizontally; average ear length: 13.8 cm (range: 11.5 to 14.1 cm).

Flock Structure

Generally, only 2 or 3 animals. A Bihar survey showed a range of 5 to 20.

Reproduction

Under farm conditions: kidding percentage on the basis of does available: 71.16%. Litter size: singles: 44.6%; twins: 51.3%; triplets: 4.1%.

13

Bionda dell'Adamello

The "Bionda dell' Adamello" is a local goat from the northern Italian Region of Lombardia. This breed takes its name from the color of its hair - Bionda in Italian means Fair - and from the mountain "Adamello" which is part of the Italian Alps. They belong to the goat population called "Alpine," from which, for years, many breeds have originated, and today still live in the European Alpine regions.

At the beginning of the century this breed was the most widespread in the flocks of the Val Camonica-Region Lombardia, district of Brescia, Northern Italy-as many photos and oral witnesses confirm its presence. Today, the "Bionda" is still bred principally in this valley and as a whole there are 40 groups totaling almost 1,200 goats in Lombardia.

Breeding

The Bionda dell' Adamello is usually raised in small flocks of 15 to 20 animals. Milking, for cheese production, is carried out after kids weaning. In wintertime only-from November till March-animals are kept in barns, otherwise they graze in uplands for the rest of the year.

Cheese Production

The "Fatuli" and the "Mascarpi" are two typical cheese produced with Bionda goat's milk. The first is smoked, the second is a whey-cheese.

According to the official data from the Italian Goat Association, these goats are noted for producing around 350 kilos of milk with 3.2% of fats and 2.96% of proteins. In some management, milk production can reach 500 kilos.

STANDARDS

The hair is very long and its color is light brown, very fair, the ears are white as well as the second thighs under the hock and the knee, the inside of the thighs and the abdomen. This last characteristic makes this breed different from the other world widespread race, the Toggenburg. In order to belong to the Bionda dell' Adamello breed standard, the goat must have two parallel white strips on its snout.

The most important fair is held at Edolo-in the district of Brescia, Lombardia-during November. During the pageant apart from the animals, farmers show their typical valley products.

14

Boer

The Boer is an improved indigenous breed with some infusion of European, Angora and Indian goat breeding many years ago. Several researchers agree that the indigenous populations were probably from the Namaqua Hottentots and from southward migrating Bantu tribes. The name is derived from the Dutch word "boer" meaning farmer and was probably used to distinguish the native goats from the Angora goats which were imported into South Africa during the 19th century. The present day Boer goat appeared in the early 1900's when ranchers in the Easter Cape Province started selecting for a meat type goat.

The South African registry was established in 1959. Since 1970 the Boer goat has been incorporated into the National Mutton Sheep and Goat Performance Testing Scheme making it the first goat breed involved in meat production performance testing.

The Boer goat is primarily a meat goat with several adaptations to the region in which it was developed. It is a horned breed with lop ears and showing a variety of color patterns. The Boer goat is being used very effectively in South Africa in combination with cattle due to its browsing ability and limited impact on the grass cover. Producing weaning rates in excess of 160% the Boer goat doe is a low maintenance animal that has sufficient milk to rear a kid that is early maturing. The mature Boer Goat ram weighs between 110-135 kg (~240-300 lbs) and ewes between 90 and 100 kg (~200-225 lbs).

Performance records for this breed indicate exceptional individuals are capable of average daily gains over 0.44 lb/day (200 g/day) in feedlot. More standard perform-ance would be 0.3-0.4 lbs/day (150-170 g/day). The ovulation rate for Boer goats ranges from 1 to 4 eggs/doe with an average of 1.7. A kidding rate of 200% is common for this breed. Puberty is reached early, ususally about 6 months for the males and 10-12 months for the females. The Boer goat also has an extended breeding season making possible 3 kiddings every 2 years.

The Boer goat was probably bred from the indigenous goats of the Namaqua Bushmen and the Fooku tribes, with some crossing of Indian and European bloodlines being possible. They were selected for meat rather than milk production; due to selective breeding and improvement, the Boer goat has a fast growth rate and excellent carcass qualities, making it one of the most popular breeds of meat goat in the world. Boer goats have a high resistance to disease and adapt well to hot, dry semi-deserts. U.S. production is centered in west-central Texas, particularly in and around San Angelo. The original US breeding stock came from herds located in New Zealand. Only later were they imported directly from South Africa.

Boer goats commonly have white bodies and distinctive brown heads. Like the Nubian goat, they possess long, pendulous ears. They are noted for being docile, fast growth, and having high fertility rates. Does are reported to have superior mothering skills as compared to other goats. Mature Boer bucks weigh between 110-135 kg (240-300 lb), and mature does between 90-100 kg (200-220 lb).

Meat goats are minimal care animals that are browsers by nature, preferring brush, shrubs, and broadleaf weeds rather than grass. Boer goats raised for meat production are typically raised on pastures. The main reasons for this are twofold: pastured goats are on average healthier animals than pen-raised goats; secondly, it costs far less to raise Boer goats on a diet of brush and weeds, than on bags of commercial feed. The ideal option is adequate year-round grazing with only mineral supplementation. Boer goats can be raised effectively in combination with cattle due their preference browse and the resulting limited impact on the grass cover.

Trace minerals are particularly important to good goat health, especially copper. Feed and mineral blocks mark "for sheep and goats" should be avoided. Copper in quantities suitable for goats is toxic to sheep, so dual sheep/goat feeds will almost always lack sufficient copper for goats. Copper deficiency is known to cause anemia, diarrhea, infertility, spontaneous abortions, and lethargy.

Fencing

The three main cost of any Boer goat operation are land, goats, and fencing. If goats are not overstocked then fencing is much less of an issue; however, goats are clever and stubborn and there is usually a few in every herd who dedicate themselves to mining every inch of the fence line looking for a weak link. Once it is discovered, they make the most of it, and the rest of the herd will not be far behind. Most farmers use 12.5 gauge fencing, with commercial 1047-6-12.5 (six-inch holes) being the least expensive and most common, however an electrically charged wire or three placed in between the goats and the fencing is often required to prevent escapes. More and more producers are experimenting with 8 and 10 strand four prong barb wire, with good success.

The most critical part of any meat goat operation is the selection of a herd sire. A high quality buck can produce high quality offspring even when mated with an average doe. Boer goats tend to gain weight at about the same rate as their sire, so a buck from a proven fast growing bloodline will command the highest price, as its offspring will tend to also be fast growers.

The primary market for slaughter goats is a 35-80 lb kid; kids should reach marketable size at weaning age. The kid of a proven fast-growing sire might weigh 80 pounds at 90 days, while the kid of a poor quality sire might weigh only 35 pounds at 90 days. An average quality buck will initially be less expensive to purchase; however, they can significant undermine an operations long-term profitability. Other criteria for a breeding Buck include:

- — Jaw alignment - most meat goats are raised on pasture. A goat with poor jaw alignment will be at a significant disadvantage when feeding on pasture; poor jaw alignment is not acceptable in a commercial herd sire.
- — Good feet and legs: needed to move about the pasture. Hoof rot is a common problem for goats that live in high rain areas if the hoofs are not clipped regularly.
- — Two well formed equal size testes in a single scrotum: -the main purpose of a buck is to breed does.

For breeding purposes, one buck is normally required for every 25-35 does. Under ideal conditions the ratio can be as high as one buck for every 50 does. Bucks are normally separated from the does except for when breeding is specifically intended. Often does are bred for six weeks every 8 months, resulting in three kid crops every two years.

Successful bucks must be able to survive on pasture. Pen-raised bucks will stay near their pen, while the does they are supposed to breed are out in the pasture.

Does

Does used to breed show quality goats are normally very large, as show goats are expected to be of large stature. For commercial meat production medium size does are normally preferred as they produce the same number of kids, but require less feed to do so.

As a general rule the more kids born per doe, the greater profit margins for the owner. Boer goats are polyestrous, and they reach sexual maturity at 5-9 months of age. A typical breading program is to produce 3 kid crops every 2 years; meaning the does are pregnant for 5 months, nurse their kids for 3 months, and then are rebred. Multiple births are common and a 200% kid crop is achievable in managed herds. Usually first time does will have one kid, but it is possible for them to have more. After that, they will have an average of two kids each time.

Weaning size is largely controlled by how much milk the mother produces, along with how long she allows each kid to nurse. Does weaning large kids should be kept, those weaning small to medium kids should be removed from the herd.

The presence of a buck causes does to come into estrus (heat) which lasts about 24-36 hours. The gestation period for does varies from 149 to 155 days.

Boer does are normally very good mothers requiring only minimal attention from the owner; however, this is not always true when a doe delivers her very first kid. First time moms should be supervised as the mothering instinct may not manifest itself the first time she delivers. After the first time, Boer does normally make excellent mothers.

Cross Breeding

While purebred bucks are usually preferred for breeding purposes, it is common to use crossbreed does for kid production, with the resulting offspring being 7/8 or more Boer. Common crosses are Boer x Spanish goat, Boer x Angora goat, Boer x Kiko goat, and Boer x Nubian goat.

Percentage Boer goats are very common in commercial meat herds, and with those just starting out in the Boer goat business due to ease of acquisition and affordability. Over time, percentage animals can be bred up to American purebred status. An American purebred is a Boer goat of 15/16ths Boer blood (for does) and 31/32nds blood for Boer bucks.

Bucks must be one generation of Boer breeding higher then does to achieve this status because they have the potential to spread their genetic pool much further then any single doe; a higher level of Boer blood lessens the chances of other breed qualities in the offspring. American purebreds can never be registered as fullblood, many breeders will still use a good American Purebred buck with excellent results.

Many producers still prefer purebred or fullblood bucks and does, and intentional crossbreeding is far from universal.

Show Goats

Although Boer goats raised specifically for show purposes may seem to have little in common with pastured commercial goats, the difference is superficial. They are bred to be larger than normal goats, and meet specific visual appearances, but these very characteristics are valuable genes to add to the commercial herd.

Boer goats were originally imported into the US and other countries for this very reason. Their value to ranchers lies in the improvement the addition of their unique genes can offer any breed of goats being raised for meat. Few producers could afford to maintain a herd of essentially useless animals. Show goats are bred to represent the most desirable characteristics of the Boer goat. And their main purpose is to introduce these animals to the public. It is also a method of recognizing the best of the best, although some really superior goats are not shown due to owner preference. Bucks and does that have been bred for show can be and often are used for commercial breeding stock. To show, most Boer goats have to be registered with either the ABGA, IBGA, or USBGA.

Overstocking

Once a section of pasture has reached is full capacity, adding more goats becomes counterproductive. Adding goats does not increase the owners profit margin, instead the forage resource is depleted, the goats must browse closer to the ground which leads to far more parasites, toxic plant problems increase, overall heath of the animals declines, and the goats fail to gain weight.

In addition rainfall becomes less effective because the soil has been laid bare, so what rain that does fall turns to run-off instead of being absorbed; thus drought management becomes much harder or impossible. All these factors reduce profit. It is not possible to starve a profit from livestock or rangeland.

In other breeds of goats it is a disqualifying trait to have more than two teats, Boer goats have been bred to have two or four, in order to easier nurse the higher numbers of kids they have been bred to have. Although the different breed registries do not agree 100% on other aspects, they do agree that up to two teats per side is acceptable.

15

Booted Goat

The Booted Goat was earlier spread throughout the uplands of St. Gallen, in the canton Glarus and in the bordering regions. It belongs to the breed of moutain goats. Until at least the 1920s it was purposefully bred, but in the 1980s it became nearly extinct. It was saved at the last moment by the foundation Pro Specie Rara. The current breeding region has its concentration in eastern Switzerland, with individual breeding groups in the central and western parts of the country.

The Booted Goat is a robust, vivacious, easily satisfied breed not specialized for milk production which is well suited for extensive keeping under extreme topographical and climatic conditions, most of all in moutainous regions. It is horned, of bright gray-brown to dark red-brown color with black or brown markings. Typical are the long beard hairs over the back and on the hind quarter, which are markedly darker or lighter than the rest of the coat of hair.

The most important goal is the reintroduction and expansion of the Booted Goat in its original region of distribution. For securing its existence Booted Goat breeding groups are being maintained, also in the remainder of Switzerland. This goal is supposed to be reached through the use of the Booted Goat as an agriculturally useful animal for the production of milk, meat and fleece.

In place of this production goal, the use of the Booted Goat for putting fallow ground into production or for caring for the landscape can be the main breeding and domestication goals. In these cases, however, the use of the products should always play a role. The goal of breeding is supposed to be reached with natural methods and more extensive care. The conditions for care are supposed to correspond as much as possible to the natural living environment and the behavior of the goat.

Responsibility for the maintanence and promotion of the breed has been assumed by the Booted Goat Breeders Club of Switzerland. This organization is connected to the herd book site for the endangered domestic animal breeds of the foundation Pro Specie Rara.

16

Brown Shorthair Goat

This breed originated with a native breed of un-uniform color, mostly light-brown to white, improved by crossing with German brown (Erzgebirgziege) bucks since the end of 19th century. It was recognized as an autonomous breed in 1954-55. The breed is dispersed in the northern and western mountain regions of the Czech Republic, settled by the German ethnic until 1945.

Similar to the German brown in appearance with glossy short hair, cinnamon to ferruginous or darkbrown coat and face. Markedly defined black eelstripe runs from cantle to tailtip. Belly, shank and hoof are black as well. Upright ears. Mostly (75-80%) hornless, with the pollness fixed through systematic selection.

Milk and fine in appearance, long legged. Average females weight about 120-140 pounds, height in withers 70-75 cm, heart girth 80-102 cm, males about 140-190 pounds, 75-85cm in withers and 90-110 cm in heart girth. Early mature, first bearing in 12-15 month of age.

The population numbers approximately 400 head now with 40% of animals held in 4 newly established commercial flocks. About 200 does included in performance testing reach 1950 pounds of milk per 300 days lactation. The best animals are reaching 2600 pounds, with 3.8% fat and 2.75% protein. The prolificacy is 185 % in average. Quadruplets are rare.

17

Danish Landrace

The Danish Landrace is derived from goats which have been in Denmark since ancient times. At the beginning of the 20th century, some crossing with goats from the German Harz region and Saanen took place. After the Second World War, the number of goats declined rapidly and a group of enthusiastic breeders collected some of the remaining animals. Unfortunately in the following years, many animals were used as recipients for embryo transfer of Angora or Boer goats or for crossbreeding with Boer for meat production.

In 1999, only 17 males and 30 females were registered in the herd book. Additionally about 100 males and 800 females exist. Most breeders keep small non-commercial herds for home consumption of milk, cheese and meat or for hobby purpose, with the exception of a few commercial milk producers. Continuing work is necessary to save this rare, but hardy and productive breed.

The Danish Landrace is a medium to large milking goat with an average annual milk production of ca. 800 liters. The animals are short to longhaired and develop a thick hair coat during winter, which protects them well against unfavorable weather. Most animals are horned but polled individuals also exist. There is a wide variation in coloring and the following colors are officially registered:

— Bezoar (brownish with white belly and black and white markings on face and legs)
— Black bezoar (black with white belly and white markings on face and legs)
— Harz color (chamois, brownish with black belly and black markings on face and legs)
— Black
— Blue (bluish gray)
— Pied (black and white or brown and white)
— White

Breeders of the Danish Landrace goat are organized in the: Foreningen for Danske Landracegeder

18

Don Goat

The Don breed was discovered in 1933-34 by an expedition of the All-Union Institute of Sheep and Goat Husbandry studying goats in the former Lower Volga territory. Because of its location, these goats were named "Don"; their habitat covers the basin of the Don river and its tributaries. The Don goats are of medium size; they have strong constitution, good conformation and adequate undercoat wool on the body, neck and belly. Males are larger than females and have a greater live weight. The average live weight of 3.5-4.5 year-old males is 70 kg (range 65-85 kg). The body shape is roundish and the skeleton is bigger than in females. Horns are large and variable in shape. Males back is long, straight and wider than in females. The average live weight of adult females is 36 kg (range 35-40kg). Kids weigh 2 kg at birth, 14 kg at weaning, 27 kg at the age of 1.5 years and 30 kg at 2.5 years.

Don goats are predominantly black, but they are sometimes white. Adult females have the following measurements (cm): height at withers 60.3, height at sacrum 62.1, oblique body length 64.5, chest girth behind shoulder blades 81.7, chest depth 30.7, chest width between shoulder joints 15.9 and width at hips 19.5.

Don goats have the highest wool production among all goat breeds, though with significant individual deviations. The average wool yield from females is 500 g; from bucks it is 1015 g (range 550-1600 g). The average wool content in the fleece is 79.4% (range 61.5-92.2%). As is known, in all other cashmere breeds and related varieties the guard hairs are much longer than the wool fibres. In this respect Don goats is that, compared with other breeds and varieties, the wool is coarser and not uniform in fineness. The average fineness of wool fibres, both fine and coarse together, is 22 μm.

Don goats are quite satisfactory milkers. The average yield in a 5-month lactation period is 130-140 litres. The milk is noted for high fat content - an average of 4.6% (range 3.3-8.2%). In the first two days after kidding the fat content is 7-12%.

Don goats are very prolific. An average kidding rate is 145-150 kids per 100 females. The goatskin is mainly used for low-grade footwear. The pelt of Don goats slaughtered in autumn has good fur quality determined by the structure of the fleece. They can be used as a basic material in the fur industry. In their fur quality Don goats are comparable with Ramanov sheep which produce the best sheepskin.

The major enterprise for Don goat breeding is the collective farm Svetly Put situated in Oktyabr district of Volgograd region. On 1st January 1984, this farm had 6500 goats which were characterized by a high classification, uniformity of production type and wool quality. Wool yield from males is 1160 g, maximum 1600 g; from females 560-630 g, maximum 1500 g. In 1982 the average yield per head was 639 g. The bulk of the herd are goats with dark grey wool. Fertility is high: up to 140-150 kids per 100 females. In 1982, the average was 114 kids born for each 100 females; 103 of them survived weaning. In the same year the farm produced 5,000 kg of high quality cashmere.

19

Ganjam Goat

The total goat population in the Ganjam distribution area, according to the 1972 census, was 0.448 m, of which 0.056 m adult males and 0.103 m adult females.

Climate

Average monthly temperature (°C)	Average	Range
minimum	23.0	16.6-26.9
maximum	30.3	27.4-32.2
Average monthly relative humidity (%)		
morning	77	72-82
evening	78	65-87
Annual rainfall (cm)	120.9	

Breed characteristics

Size

	Adult males	Adult females
Body weight (kg)	44.05 ± 0.13 (32)	31.87 ± 0.37 (143)
Body length (cm)	76.2 ± 1.2 (32)	67.6 ± 0.3 (143)
Height at withers (cm)	84.5 ± 0.7 (32)	77.1 ± 0.3 (143)
Chest girth (cm)	83.1 ± 0.9 (32)	74.6 ± 0.3 (143)

Conformation

Tall, leggy animals. The coat may be black, white, brown or spotted, but black predominates. Hairs are short and lustrous. Ears are mediumsized; ear length: 14.50 ± 0.15 cm. Both sexes have long, straight horns, directed upward; horn length: 20.97 ± 0.33 cm. Tail is medium-long.

Flock Structure

The average flock contains 84.5 ± 37.5 individuals, of which 2.9 adult males, 59.8 adult females and 21.8 young.

Reproduction

In farmers' flocks: kidding percentage: 82.0 ± 4.2%; litter size: singles: 98.4%; twins: 1.6% . Under farm conditions: age at first kidding: 1 066.9 ± 25.6 days; kidding interval: 376.1 ± 6.7 days; service period: 216.8 ± 5.1 days.

20

Gohilwadi Goat

The total goat population in the Gohilwadi distribution area, according to the 1972 census, was 0.369 m, of which 0.017 m adult males and 0.277 m adult remales.

Climate.

Average monthly temperature (°C)	Average	Range
minimum	21.0	11.1-27.1
maximum	33.4	27.6-39.6
Average monthly relative humidity (%)		
morning	56	39-77
evening	39	21-64
Annual rainfall (cm)		60.01

Breed characteristics

Size

	Adult males	Adult females
Body weight (kg)	37.10 ± 1.42 (22)	36.03 ± 0.38 (173)
Body length (cm)	73.4 ± 1.3 (22)	72.4 ± 0.3 (173)
Height at withers (cm)	81.2 ± 1.2 (22)	79.5 ± 0.3 (173)
Chest girth (cm)	74.7 ± 0.9 (22)	75.2 ± 0.2 (173)

Conformation

Large animals. The coat is black and covered with coarse long hairs. Nose line is slightly convex. Ears are tubular and drooping; ear length: 17.5 + 2.7 cm. Both sexes have slightly twisted horns, turned backward; horn length: 12.5 ± 0.20 cm.

Flock structure

The average flock contains 24.2 ± 2.1 individuals, of which 0.7 adult males, 17.1 adult females and 6.4 young.

Reproduction

In farmers' flocks: kidding percentage: 72.9 ± 6.2; litter size: singles: 84.2 ± 4.1%; twins: 15.8 ± 4.1%.

21

Goinge Goat

The Goinge goat is not breed for how much milk or meat they produce. The breeding selection is quite random exccept for the obvious effort to not breed animals who are to closely related. This however is not an easy task since all Goinge goats come from two pregnant goats found in the area around Tyringe in the south of Sweden. "The Society of the the Peseantry Goat" or in swedish: "Föreningen Allmogegeten" have at Gene Bank so they can controll the breeding of the Goinge Goat.

The Goinge Goat originates from a type of goats that were common in all over the south of sweden in the 19:th century.

This goat is usually horned and is seen in colorations of gray, brown, white or pied.The hair on this breed is usually long. This is a quite robust goat.

22

Golden Guernsey

The Golden Guernsey is a rare breed which originated in the Channel Islands off the coast of Britain. The breed was developed from local breeds mated to Anglo-Nubian and Swiss breeds during a period from 1920 to 1950.

Golden Guernseys are kept for their milk and are recognized as a dairy breed by the British Goat Society. The production levels are lower than what is commonly found in the Swiss breeds but the milk is higher in butter fat and proteins which make it an excellent choice for cheese making.

The breed is used primarily for milk production. Its coloration varies from cream to brown and the breed is usually hornless (polled). The Golden Guernsey was used in the development of the English Guernsey.

This attractive animal from the Bailiwick of Guernsey in the Channel Islands, has grown in popularity as a useful household goat with a pleasant temperament and steady yield.

Its origin is uncertain though golden goats have been in the Bailiwick for many years. It is not certain how long goats have been on Guernsey, but fine boned skeletons of goats were found close to the Dolmens, megalithic burial grounds some 2000 BC.

The first documented evidence is found in an old guidebook to the island in 1826 when the golden cow, golden goat and golden donkey were all mentioned. At this time it was common practice for ships to carry goats on sea journeys to provide fresh milk for passengers. When they reached a port these goats were put ashore.

It is thought that Golden Guernseys originated from Maltese goats descended from various breeds including Chamoisee and Syrian, the latter being golden coloured with tipped up ears. However, the Maltese goats were crossed with local Channel Island goats and occasionally a golden kid would crop up. It was these goats of unknown ancestry who qualified for the early Register on account of their colour. Registration on the Island first started in 1922 in the Guernsey Goat Society's general herd book.

There were always golden goats on the Island with several breeders developing and maintaining the breed. During the German occupation, although the GGS disbanded for the 5 year period, Golden Guernseys were still being registered.

One of the most well known ones was Miss Milbourne of L'Ancresse, Guernsey, who kept a large herd (50+) goats that browsed on L'Ancresse common (now a golf course!) and played an important part in reviving the breed in the 1930's, encouraged by Dr. Tracey, a British Goat Society Judge. She managed to retain her flock during the occupation and on her death her animals passed into the care of a trust - now disbanded.

In 1965 the Guernsey Goat Society opened a special section of the herd book for the Golden Guernsey as they were breeding true to type. In 1972 the Golden Guernsey Goat Association was started, now known as the Golden Guernsey Goat Breed Society. The objectives of the Society are to encourage the keeping of Golden Guernsey Goats in the Bailiwick of Guernsey, to retain the purity of this unique breed, to lay down rules and regulations and eliminate incipient bad points, to improve the milk yield, conformation and longevity factors, to provide an Island based authority on this local breed, to advise owners and breeders throughout the world.

The Golden Guernsey Goat is distinguished by its golden skin and hair, which can be of some length along the back and down the hind legs. There is some variation in golden colour allowed and small white markings and blaize or star on head, but no Swiss markings. It is fine boned with a straight or slightly dished facial line, the ears have a slight upturn at the tip and the neck is slender. It is a fairly small goat, males weighing 190-200 lbs and females 120-130 lbs. Conformation is neat and pleasing.

The milk yield is steady with good butterfats. Its reliability in food conversion and placid temperament make it an economic proposition as a family animal.

British Isles

In 1965 the first goats were exported from the Bailiwick to breeders on the mainland and have continued over the years. The English Golden Guernsey Club was formed, now known as the Golden Guernsey Goat Society. In 1970 the British Goat Society opened a Breed Register for the Golden Guernsey in its Herd Book.

In 1973 The Rare Breeds Survival Trust was established and the Golden Guernsey was and still is included on their Priority List in Category 3 (Vulnerable).

The dangers that especially threaten minority breeds are the narrowing of their genetic base and the problems of inbreeding. These have been particularly evident among Golden Guernsey goats, although significant progress has been made.

It must be remembered that before 1980 all Golden Guernseys on the mainland were descended from only 108 imports and of this number 30 failed to breed, so to be more correct prior to 1980 all Goldens on the mainland were descended from only 68 imports. Although there have been some imports since then this has not been in significant numbers.

1980 saw the breed awarded its first Challenge Certificate when Duesquene GG150H was made Best in Show at Devon Male and Youngstock Show.

Novington Boris became the first Sire of Merit in 1983, rapidly followed by Sark Roc and Chickering Pepetidor. Since then there have been a further 8 Sire of Merit awards within the breed, making a total of 11.

It was not until 1987 that the British Goat Society finally recognised the Golden Guernsey as a breed by allowing them to compete for Breed Challenge Certificates at shows. Only one year after this the breed gained its first Breed Champion in R105 Acorus Cordelia*Br Ch. To date (2003) there have been 18 male Golden GuernseyBreed Champions and 11 female Breed Champions. In Herd Book 122 (December 1995 to Nov 1996) the British Goat Society replaced the Golden Guernsey Register with the Golden Guernsey Section and finally, in 2001, changed their regulations so that the Golden Guernsey was on a parity with all other breeds of goats in the Herd Book.

23

Grison Striped Goat

The Grisons Striped goat is a robust breed which feels at home most of all in the moutains. It prefers stony and steep landscapes.

In the 1930s it achieved Honor for Recognition and Promotion, in contrast to the other goats from the canton Graubünden. The strength of the breed is its suitability for mountainous regions. Its milk production, however, is not equal to that of the milk breeds.

Low milk production is a reason why the stocks in the last decades have sunk continually. The remaining breeders had to recognize that the breed was becoming more endangered the longer it existed.

In the fall of 1992 twelve bucks registered in the herd books remained, a dangerously narrow basis. With the support of the canton the breeders have joined themselves together into a Community of Interests of the Grisons Striped Goat. They intend to take the promotion of the breed by the hand and above all devote attention to milk production and the avoidance of inbreeding. In 1994 four bucks and two nannies were imported from Great Britain which have the same color characteristics as the Ray Goat, but significantly better milk production.

The friends of the Grisons Striped Goat have also had to devote themselves to the widespread problem of CAE, a reason for the dramatic decline of the Ray Goat. Nearly all associations as well as private individuals have participated in the restoration, so that once again an upward trend is noticeable.

24

Jamnapari Goat

The name is derived from the location of the breed beyond the river Jamna (Jamna Par) in Uttar Pradesh. Agra, Mathura and Etawa districts in Uttar Pradesh and Bhind and Morena districts in Madhya Pradesh. However, the pure stocks are found only in about 80 villages in the vicinity of Batpura and Chakar Nagar in Etawa district.

The total goat population in the Jamnapari distribution area, according to the 1972 census, was 0.58 m. However, officials of the Animal Husbandry Department of Uttar Pradesh state that the total number of pure-bred Jamnapari does not exceed 5 000; these are located mostly in the Chakar Nagar area, between the Jamna and Chambal ravines. There is a serious need for conservation, multiplication and further improvement of the breed, considering the extremely small numbers of pure-bred animals remaining.

Jamnapari is one of the largest goats in India; it has been extensively utilized to upgrade indigenous breeds for meat and milk, and has been taken to neighbouring countries for the same purpose.

Climate

Average monthly temperature (°C)	Average	Range
minimum	19.0	7.4-29.5
maximum	32.3	22.2-41.8

Average monthly relative humidity (%)		
morning	60	30-81
evening	41	18-75
Annual rainfall (cm)		76.54

Breed Characteristics

Size

	Adult male	Adult female
Body weight (kg)	44.66 ± 1.89 (49)	38.03 ± 0.63 (168)
Body length (cm)	77.37 ± 1.23 (49)	75.15 ± 0.46 (168)
Height at withers (cm)	78.17 ± 1.25 (49)	75.20 ± 0.38 (168)
Chest girth (cm)	79.52 ± 1.2 (49)	76.11 ± 0.38 (168)

Conformation

Large animals. There is a great variation in coat colour, but the typical coat is white with small tan patches on head and neck. The typical character of the breed is a highly convex nose line with a tuft of hair, yielding a parrotmouth appearance. The ears are very long, flat and drooping; ear length: 26.79 ± 0.21 cm. Both sexes are horned; horn length: 8.69 ± 0.27 cm. Tail is thin and short. A thick growth of hair on the buttocks, known as feathers, obscures the udder when observed from behind. The udder is well developed, round, with large conical teats.

Flock Structure

The average flock contains 16.0 ± 2.57 individuals, of which 0.25 adult males, 8.65 adult females and 7.1 young.

Management and Feeding

The flocks, stationary, are maintained primarily on tree loppings, acacia pods and stubble of cultivated crops. Some supplementary feeding (200 to 250 g/day of Arhar and gram) is given to lactating animals. The animals are housed in the courtyard within the living quarters, under a thatched shed with a thorny fence. Most animals are docked. Castration does not take place.

Reproduction

In farmers' flocks: kidding percentage: 124.25 ± 6.05; litter size: singles: 69.0 ± 7.1%; twins: 27.8 ± 6.46%; triplets: 3.2 ± 0.0%. Under farm conditions: age at first kidding: 737.0 ± 21.25 days; kidding interval: 229.3 ± 26.71 days; service period: 101.4 ± 13.03 days; kidding percentage: 79.6; litter size: singles: 56.2%; twins: 43.1%; triplets: 0.7% .

Mortality

In farmers' flocks: young: 5.75 ± 2.5%; adults: 4.21 ± 1.97%.

Breeding

Flocks are pure-bred. Selection in bucks is based on dam's milk yield.

PERFORMANCE

Milk: Average lactation yield: 201.96 ± 6.65 kg; lactation length: 191 ± 6 days; average daily yield: 1.752 ± 0.031 kg; dry period: 115 ± 3 days. In farmers' flocks, average daily yield: 2.15 ± 0.30 kg; average lactation length: 255 ± 6.7 days.

Meat: body weight (kg)

At birth	4.27 ± 0.04 (372)
3 months	12.11 ± 0.45 (13)
6 months	15.56 ± 1.67 (5)
9 months	24.00 ± 1.16 (5)
12 months	29.65 (100)

Dressing percentage on pre-slaughter live-weight basis: 44.57 ± 1.09 at 6 months, and 48.16 ± 0.91 at 9 months. Bone/meat ratio: 1:3.89.

25

Jhakrana Goat

Jhakrana and a few surrounding villages near Behror, in the Alwar district of Rajasthan. The number of animals of this breed is rather small, as it is restricted to a very limited area. Considering that it is a good indigenous dairy breed, there is need for its conservation.

Jhakrana goat are large animals. The coat, predominantly black with white spots on ears and muzzle, is short and lustrous. Face line is straight. Forehead is narrow and slightly bulging. The breed is quite similar to Beetal, the major difference being that Jhakrana is longer. Ear length is medium: 13.8 ± 0.19 cm. Udder is large, with large conical teats.

Flock structure

The average flock consists of 14.2 ± 2.1 individuals; including 0.05 adult males, 9.08 adult females and 5.07 young.

Climate

Average monthly temperature (°C)	Average	Range
minimum	18.4	8.3-27.3
maximum	31.7	22.0-40.6
Average monthly relative humidity (%)		
morning	53	29-68
evening	37	18-71
Annual rainfall (cm)		64.81

Breed Characteristics

Size

	Adult males	Adult females
Body weight (kg)	57.80 ± 3.50 (10)	44.48 ± 0.52 (178)
Body length (cm)	84.10 ± 2.11 (10)	77.74 ± 0.36 (178)
Height at withers (cm)	90.40 ± 1.61 (10)	79.12 ± 0.29 (178)
Chest girth (cm)	86.00 ± 1.91 (10)	79.13 ± 0.31 (178)

Reproduction

In farmers' flocks: kidding percentage: 141.8 ± 10.05. Litter size: singles: 57.0 ± 7.6%; twins: 41.0 ± 7.5%; triplets: 2.0 ± 1.2%.

Breeding

Pure breeding. Males are selected on the basis of their dam's milk yield. Breeding bucks are generally produced from within the flock. Owners of small flocks who do not maintain a buck utilize services of the buck from a neighbour's flock.

Performance

Milk: In farmers' flocks, average daily milk yield: 3.18 ± 0.17 kg; lactation length: 231.8 ± 6.3 days. Under farm conditions, lactation yield: 121.80 ± 8.8 kg; lactation length: 114.7 ± 18.5 days; daily milk yield: 0.988 ± 0.06 kg.

26

Jining Grey

The Jining Grey of Shandong Province, is noted for the attractive wavy patterns of its kid-pelt, which is the traditional commodity in internatinal markets. Does reased in temperate areas and with abundant feed reach sexual maturity at 3-4 months of age, and may give two kiddings a year, or three kiddings in two years. They are very prolific, with a average kidding rate of 294 percent.

Both sexes are horned and carry a forelock. The color patterns vary between black, white or black and white. They are a small bodied goat with the males weighing approximately 33.2 kg (~75 lbs) and the females 25.4 kg (~56 lbs).

The Jining Grey also has some production of cashmere. The production from the males will range from 50-150 g with cashmere fiber being 18-30 percent of the total fleece. Production in the females is lower, ranging from 25-50 g and cashmere comprizing 16-20 percent of the fleece. The fiber diameter in both sexes averages 13.0 microns.

27

Kiko Goat

The appropriately named Kiko goat was purpose-bred in New Zealand for meat production - the Maori word "kiko" meaning flesh or meat. The developers of the breed were Garrick and Anne Batten of Nelson in the northern South Island, and they developed the Kiko from feral goats that had been liberated or had escaped over the last hundred years or so of European settlement. These feral goats could be found in many places throughout the country, and although they were hardy, they were relatively small and produced little meat or milk.

The Kiko breed was established by crossbreeding selected feral does with Anglo-Nubian, Toggenburg and Saanen bucks, with further cross-breeding in the second and third generations. After four generations of selective breeding - selection being on the grounds of survivability and growth rate in a hill country environment - a dramatic improvement in liveweight and animal performance was achieved. By 1986 the Kiko breed was established and the herd was closed to further cross-breeding.

Within New Zealand, control of the breed has remained with the original developers. However, Kikos were exported to the United States in the 1990s, and there are now a number of enthusiastic breeders of Kikos in that country.

To develop the breed, an unselected population of several thousand feral goats were captured and subjected to stringent selection for specified traits. Does were bred to domestic purebred bucks to introduced superior genetics. The goal was to increase the frame size, muscling and milk. A very strict breeding program was followed. Geneticists in the Ministry of Agriculture and Fisheries assisted in drawing up guidelines and offering advice for the program.

Overall the two main criteria for selection were survivability and weight gain. No shelter was provided. No assistance was given during kidding. No supplem-ental feeding was offered in range conditions generally considered demanding. No hoofs were trimmed. Minimal parasite control was administered.

The animals that performed exceptionally were retained, and in 1986 the herd was closed to outside bucks. Outside does were slowly added if they met the selection criteria.

Today the Kiko is a large framed, early maturing goat that demonstrates exceptional conversion rates. They are not as heavily boned as some breeds; consequentially, they exhibit high cutability. This trait, combined with their lean, succulent carcass, yields a product that puts money back into the producer's pocket.

In circumstances that would stress the best of the four-legged critters, the Kiko prevails. The advantage to the commercial rancher is obvious. Unlike the hobby rancher who may have more time and resources than goats, the commercial rancher does not. Pet names tend to go out the window, and the burden of survival rests largely on the animal. It is not that ranchers get hard-hearted. They are still willing to get up in the middle of the night and often put care of their charges before their own well-being, but physical limitations draw the line.

The hybrid vigor resulting from crossing any breed of goat with the Kiko is astounding. In 1994, Goatex Limited performance-tested a group of Kiko/Boer cross males. The most startling aspect was that the 90-day weaning weight for the crossbred males was 1.86 kg greater than for purebred Boers and 1.71 kg higher than for purebred Kikos. The experiment was repeated in 1995 and the results were again confirmed.

In cutting trials, the Kiko-cross carcass yielded more meat per carcass than a purebred Boer but slightly less than a purebred Kiko. For the commercial meat goat rancher this appears to be the premier cross. Kikos come in every color. However most Kikos in North America are white because mostly white goats were imported, and the white coat color is dominant.

Interestingly, some breeders in New Zealand and South Africa believe that growth rates in goats are related to coat color. There is some anecdotal evidence to support this belief. As the Kiko Breeding Up Program expands, more colored genes will be introduced into the gene pool with the resulting consequence of more colored Kikos. Some people believe this will be an asset to those ranchers living in areas with heavy predator infestation.

In order to protect the integrity of the breed, Goatex Group Ltd. formed a New Zealand Kiko Registry. All purebred animals leaving the country were fitted with a microchip and safeguards were built into the recordkeeping system to track every Kiko leaving New Zealand.

A limited number of registered purebreds have been brought into the United States. The American Kiko Goat

Association is the American liaison with the New Zealand Registry for registering percentage and purebred Kiko Goats. A percentage breed-up program is available to new breeders with the 15/16 female (4th cross) considered purebred.

28

Kinder Goat

The Kinder goat, as it has evolved, is a joy to milk and an ideal small homestead milk producer. Though smaller, Kinder goats are required to meet the same standards of production to be eligible for star milker awards as their larger counterparts registered by the American Dairy Goat Association. With a base production of 1,500 pounds of milk and/or 52.5 pounds of butterfat in 305 days or less, freshening at 2 years or less in age, these hard working little animals stand up well in comparison. The original, Zederkamm Liberty, has a 305-day record starting at 3 years and 6 months of age of 1,730 pounds of milk and 115 pounds of butterfat.

Since the breed is being developed as a dual purpose breed, it is also important to note that the wethers (does as well) make extremely desirable meat animals. Usually born weighing 4 or 5 pounds, they grow rapidly at a rate of about 7 pounds per month. Recently some 6-month-old wethers were slaughtered weighing about 50 pounds and dressing out at 30 pounds. This makes the dressing percentage a very favorable 60 percent. Older wethers have an even higher dressing percentage with some 14-month-old wethers having a live weight of 80 pounds and a carcass weight of 50 pounds - a dressing percentage of nearly 63 percent.

It is easily possible for a Kinder doe weighing about 115 pounds to produce five kids who in 14 months can weigh 80 pounds each and dress out at 50 pounds thereby producing 250 pounds of meat each year.

29

Kutchi Goat

The total goat population in the Kutch district, according to the 1972 census, was 0.402 m, of which 0.018 m adult males and 0.298 m adult females. Kutchi goats are large animals. The coat is predominantly black, but a few white, brown and spotted animals are also found. The hair is coarse and long. The nose is slightly Roman. The ears are long, broad and drooping; ear length: 22.0 ± 0.26 cm. Both sexes have short, thick horns, pointed upward; horn length: 10.6 ± 0.19 cm. The udder is reasonably well developed; teats are conical.

Flock Structure

The average flock contains 83.3 ± 14.5 individuals, of which 2.4 adult males, 70.3 adult females and 10.6 young.

Reproduction

In farmers' flocks: kidding percentage: 75.5 ± 4.5; litter size: singles: 84.1 ± 4.2%; twins: 15.2 ± 4.2%; triplets: 0.7 ± 0.4%.

Climate

Average monthly temperature (°C)	Average	Range
minimum	20.1	10.1-27.4
maximum	33.0	26.1-38.7
Average monthly relative humidity (%)		
morning	63	45-80
evening	36	17-64
Annual rainfall (cm)		34.87

Breed Characteristics

Size

	Adult males	Adult females
Body weight (kg)	43.50 ± 1.16 (24)	39.29 ± 0.40 (176)
Body length (cm)	77.1 ± 1.3 (24)	75.0 ± 0.3 (176)
Height at withers (cm)	86.4 ± 0.7 (24)	82.4 ± 0.3 (176)
Chest girth (cm)	78.4 ± 0.7 (24)	76.1 ± 0.2 (176)

Mortality

In farmers' flocks: young: 15.0 ± 2.4%; adults: 5.5 ± 1.2%.

Performance in Farmers' Flocks

Milk: Average yield: 1.84 ± 0.15 kg ; average lactation length: 117.7 ± 7.3 days.

Hair: Average yield per clip: 229.3 ± 23.9 g . The annual hair yield may be taken to be approximately twice this figure, since the animals are shorn twice a year.

30

LaMancha

The LaMancha goat originated in Oregon by Mrs. Eula Frey from short-eared goats of a type found not only in LaMancha, but throughout spain. It has excellent dairy temperament and is an all-around sturdy animal that can withstand a great deal of hardship and still produce. Through official testing this breed has established itself in milk production with high butterfat.

The LaMancha face is straight with the ears being the distinctive breed characteristic. There are two types of LaMancha ears. In does one type of ear has no advantage over the other.

— The "gopher ear" is described as follows: an approximate maximum length of one inch but preferably non-existent and with very little or no cartilage. The end of the ear must be turned up or down. This is the only type of ear which will make buck eligible for registration.

— The "elf ear" is described as follows: an approximate maximum length of two inches is allowed, the end of the ear must be turned up or turned down and cartilage shaping the small ear is allowed.

Any color or combination of colors is acceptable with no preferences. The hair is short, fine and glossy.

LaManchas do have ears; that is to say an auditory canal and other internal structures. It is the pinna that is visibly shrunken. It is not usually possible to use ear tags or ear tattoos for identification of individual LaMancha goats. Tattoos are normally placed on the tail web.

The rather small ears on LaManchas make them a unique breed. However, it is possible to breed the small ear 'out of them'. The reason only gopher eared bucks are allowed to be registered is because two 'elf' eared animals bred together can create an erect (normal) eared kid. Some breeders prefer the gopher ears on all of their animals, but the type of ear is purely cosmetic.

31

Marwari Goat

Marwar region of Rajasthan, comprising Jodhpur, Pali, Nagaur, Bikaner, Jalore, Jaisalmer and Barmer districts. The breed also extends into certain areas of Gujarat, especially Mehsana district. The total goat population in the Marwari distribution area, according to the 1972 census, was 3.914 m, of which 0.072 m adult males and 2.484 m adult females.

Climate

Average monthly temperature (°C)	Average	Range
minimum	20	8.2-28.3
maximum	33.7	23.8-41.8
Average monthly relative humidity (%)		
morning	54.6	35.3-78
evening	32.3	18.3-57.6
Annual rainfall (cm)		33.17

Breed Characteristics

Size

	Adult males	Adult females
Body weight (kg)	33.18 ± 1.77 (17)	25.85 ± 0.29 (404)
Body length (cm)	70.97 ± 1.65 (17)	63.51 ± 0.44 (404)
Height at withers (cm)	74.74 ± 1.61 (17)	69.29 ± 0.22 (404)
Chest girth (cm)	71.68 ± 1.41 (17)	68.60 ± 0.26 (404)

Conformation

Medium-sized animals. Predominantly black with long shaggy hair coat. In about 5% of individuals, white or brown patches are also observed. Beard is present in both sexes. Ears are flat, medium in length and drooping; ear length: 16.38 ± 0.20 cm. Both sexes have short, pointed horns, directed upward and backward; horn length: 10.10 ± 0.18 cm. Tail is small and thin. Udder is small and round, with small teats placed laterally.

Flock Structure

Average flock size: 48.0 ± 4.6, containing 0.6 adult males, 36.4 adult females and 11.0 young.

Reproduction

In farmers' flocks, kidding percentage: 52.5 ± 3.4; litter size: generally single.

Mortality

Young: 23.1 ± 3.2%; adults: 9.6 ± 1.5.

Breeding

Pure breeding. Breeding males are selected on size and hair production, mostly from within the flock. The Animal Husbandry Department of the Government of Rajasthan has introduced cross-breeding with Jamnapari to increase milk production.

Performance

Milk: Under farmers' flock conditions: average daily milk yield: 0.530 ± 0.32 kg; lactation length: 196.3 ± 14.2 days. Under farm conditions: average lactation yield: 91.39 ± 6.8 kg; length: 105.80 ± 14 days; daily milk yield: 0.713 ± 0.047 kg.

32

Malawi Goat

Malawi - whose background information has been described previously has a goat population of 631000, second only to that of cattle at 908000. The flock growth rate is 1.8 percent annually. An estimated 30 percent annual offtake, 25 to 30 kg live weight and a killing out of 57 percent gives 2700 to 3000 tonnes of meat annually, showing a low consumption of goat meat alone of 0.42 to 0.46 kg per head per year, with no change since 1960. Of all meat from domestic ruminants consumed, goat meat represents 11 percent.

Constraints to goat production in Malawi - apart from the communal landownership system - are the low genetic potential of the animals, problems of providing improved husbandry and management, the absence of marketing policies, poor attitudes towards goat keeping and inappropriate extension strategies for goat keeping, only to mention a few. In 1971, a small flock of pure-bred Boer goats was established at Mikolongwe in southern Malawi for crossing with local Malawi goats in order to improve meat production in the country. A similar flock is available at the Bunda College of Agriculture, University of Malawi, for the same purpose. In 1983, the Malawian Government, in collaboration with the Government of the Federal Republic of Germany, established a goat breeding ranch in Salima, central Malawi. Managed by the Malawi-German Livestock Development Programme (MGLDP), cross-breeding as well as selection programmes are currently under way there.

For any improvement of goat production to benefit the rural community and smallholder farmers through research and policy-making, the physical characteristics, reproductive ability, feeding systems, productivity and health aspects of goats must be understood. To this effect, some work has been done by the Ministry of Agriculture and the University of Malawi, but no attempt has been made to unite all of the available information.

The majority of the goat population in Malawi is local. The goats apparently came from Asia and spread southwards from Egypt and Ethiopia down to Uganda, Kenya, the United Republic of Tanzania, Rwanda and Burundi and then to Malawi. As a result, the Malawi goat bears some resemblance to the small East African type.

At the same time, Joubert grouped the local goats together with those from Zambia, Zimbabwe and Botswana. This suggests that some of the goats may have been brought together with cattle as the Zulus moved up from the south, though this needs to be ascertained. Nonetheless, the Malawi goats have developed through natural selection and adaptation to harsh environments, utilizing limited vegetation in long dry seasons.

Ninety percent of the indigenous goats are kept by smallholders in flocks of up to 20, the average being six animals. Collections of local goats are obviously also kept at government farms. The cattle:goats:sheep ratio of 10:1:0.2 indicates that cattle are more important than small ruminants and that goats are far more important than sheep. About 42 percent of the goats are found in the semi-arid low-lying areas of the Lower Shire Valley and the Lakeshores, follcwed by plateau areas (39 percent) and then high altitude areas (19 percent). No work has been conducted so far to characterize the goats and the inherent production systems found in these different ecological zones.

Physical Characteristics

The local Malawi goats resemble the small East African goat

breed. The goats are hardy and small- to medium-sized and the hair is short.

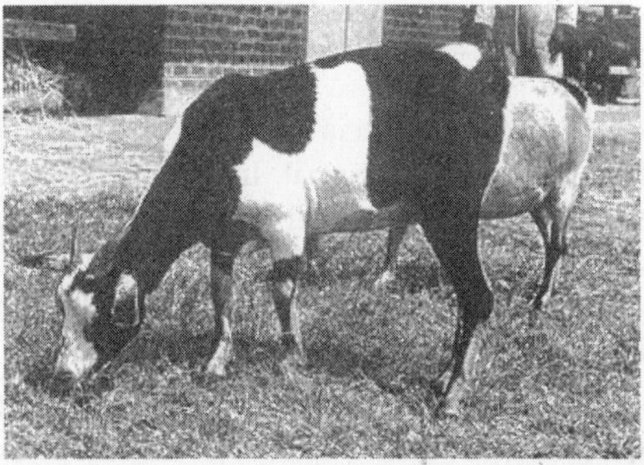

In many cases, horns are present, pointing backwards. Beard and sometimes wattles are seen, and ears are mostly short and prick-type. They have a wide range of colours, including black, black-and-tan, red-and-white and grey. Sometimes they look like deer, with a black line of hair extending down the back and up to the tail, as is common with deer-like goats. Face patterns are variable.

Weight and Size

The birth weight of the kids is influenced by breed, age of the dam, season of birth, sex and litter size. Single-birth kids weigh 0.4 to 0.5 kg more than multiple-birth kids. Weaning weights are lower in female than in male kids by 3.4 to 8 kg. At birth the weight of multiple-birth kids is 80 percent, and at weaning 69 percent, of the weight of single-birth kids. This demonstrates that multiple-birth kids grow slightly faster than single-birth kids, although single-birth kids still show a weight advantage throughout their lives.

Kids born in the early part of the dry season (April/May) tend to be heavier than those born later in the year (November/December), probably because of the availability of browse shrubs and trees, including field crop residues, coupled with lower parasitic infestation during the dry season. Mature weight of local Malawi goats is about 21 to 28 kg, with a body length of 61 cm and a heart girth of 66 cm. These are close to the values given by Devendra and McLeroy.

Reproduction

Environmental factors, especially the photoperiod, can affect the reproductive cycle of the goat. In the tropics, the variation in day length is less marked and so goats will breed throughout the year. Interactions of tempera-ture and nutrition may be important in determining the pattern of reproduction in the tropics.

Kanyerere and Kamwanja found a significant difference in age at first kidding between locals (20 months) and local x Boer crosses (25 months), while Karua found that the age at first kidding for locals was 16 to 18 months. Fertility at first oestrus was 44 percent for locals and 45 percent for local x Boer crosses, while the age at first observable heat was 15 months for locals and 16.5 months for local x Boer crosses at Bunda College. According to Reynolds, the Malawi goat has a kidding percentage of 85 to 126 percent, the range reflecting deaths of foetuses and kids.

These results are in agreement with those reported by Karua. Mean gestation length in local goats was reported to be 148 days. The incidence of multiple kidding is approximately 15 to 20 percent. The interval from kidding to service is 18 to 19 weeks. This postpartum anoestrus plus a normal gestation period of 21 weeks shows that the kidding interval ranges between 39 and 40 weeks. Kasowanjete, Stotz and Zerfas and Karua observed a kidding interval of 45 weeks in local goats on smallholder farms in Salima, central Malawi.

On station at Bunda College, the kidding interval was 47 weeks for locals and 43 weeks for local x Boer crosses. These values obtained for local Malawi goats are not different from those quoted by Devendra and McLeroy. This shows that local Malawi goats kid four times in three years. This is affected mainly by the anoestrus period, which is rather long.

It seems that the only apparent anoestrus period occurs towards the end of the dry season, from mid-August through October, when crop residues in harvested fields are exhausted; hence, goats do not kid as much in February and March.

FEEDING SYSTEMS

The local Malawi goats, like many other goats, are inquisitive animals and they can quickly cover long distances in search of food. Browse is preferred by the Malawi goats and they are able to select the types of trees, shrubs and also grasses they prefer. Three feeding systems are normally observed in Malawi. The first is the village system, where the goats are tethered during the rainy season to prevent crop destruction in fields and then released during the dry season into harvested fields. The goats lose weight during the rainy season since, tethered, their grazing is restricted. In the dry season, they gain weight as they are free to graze on the crop residues and grain remnants available in the harvested fields.

In the second system, which is a modification of the first, the goats are tethered in the morning hours and then herded by young boys after school. The third system is the extensive system, where goats browse, graze and scavenge on the feeds available in areas where there are no crops. The goats kept under this system are usually managed by young boys who either belong to the household or are employed.

Seasonal fluctuations in feed supply and quality and infestations with internal parasites (helminths) affect body weight changes and mortality, especially of kids. In all systems, crop residues are sometimes utilized when crops have been harvested.

33

Mehsana Goat

The total goat population in the Mehsana distribution area, according to the 1972 census, was 0.736 m, of which 0.030 m adult males and 0.530 m adult females.

Mehsana goats are large animals. The coat is black, with white spots at the base of the ear. Nose line is straight. The hair coat is long and shaggy. Ears are white, leaf-like and drooping; ear length: 15.8 ± 0.55 cm. Both sexes have slightly twisted horns, curved upward and backward; horn length: 11.0 ± 0.20 cm. The udder is well developed; the teats are large and conical.

Flock Structure

The average flock contains 54.7 ± 13.1 individuals, of which 0.7 adult males, 39.4 adult females and 14.6 young.

Reproduction

In farmers' flocks: kidding percentage: 69.5 ± 7.1; litter size: singles: 89.7 ± 2.7%; twins: 10.3 ± 2.7%.

Mortality

In farmers' flocks: young: 13.0 ± 0.35%; adults: 3.4 ± 1.1%.

Climate

Average monthly temperature (°C)	Average	Range
minimum	20.5	11.9-27.4
maximum	34.2	28.7-40.7
Average monthly relative humidity (%)		
morning	64	47-86
evening	37	18-69
Annual rainfall (cm)	82.3	

Breed characteristics

Size

	Adult male	Adult female
Body weight (kg)	37.14 ± 1.51 (11)	32.39 ± 0.38 (185)
Body length (cm)	71.2 ± 1.0 (11)	68.0 ± 0.3 (183)
Height at withers (cm)	80.4 ± 1.2 (11)	74.3 ± 0.2 (188)
Chest girth (cm)	76.9 ± 1.2 (11)	73.0 ± 0.3 (188)

Breeding

Pure breeding. Males are selected primarily on body size.

Performance

Milk: Average daily yield: 1.323 ± 0.134 kg; average lactation length: 197.2 ± 7.5 days.

Hair: Average yield per year: 210.2 ± 20.4 g.

34

Murcia-Granada

The Murcia-Granada goat is a native breed originated in the southeast of Spain. Some years ago it was introduced to Argentina, Brazil, Mexico and Venezuela. Animal have also been exported to North Africa. In Spain it has been selected for rusticity, high performance and for its excellent milk production.

The Murcia-Granada has a small-medium size with a liveweight near 30-50 kg for females and 50-60 kg for males. It has a uniform black to brown color skin and hair. Females have short and thin hair but longer and stronger in males. Males have usually pronounced beard and are horned. The ears are of medium size and erect. The tail is short and erect. The udders are large, well-shaped and with well-placed teats. The extremities are thin but solid and of medium size.

Types of Murcia-Granada

"Veguesi" from irrigated and flat land, and "Montana" from mountains areas. Veguesi goat is better milk producer than Montana and is larger, coloration is similar on both varieties.

The Murcia-Granada is well adapted to the hot and dry conditions of the semiarid areas of southeastern Spain. It is the most productive domestic animal in this climate because of its ability to maintain a high milk production under less than ideal conditions.

The Murcia-Granada flocks graze adverse terrain and feed on the by-products of the agroalimentary industry with a very good transformation index. This breed produces near 500 kg of milk in 280 days of lactation, with a composition of fat (5.3%) and protein (3.4%) better than other breeds in Mediterranean areas. Most of milk production in Spain is used for cheese production.

35

Mytonic Goat

Myotonic goats are also called Wooden Leg goats, "stiff leg" or Tennessee fainting goats. These are one of the few goats that are indigenous to the U.S. There are two strains of this animal. Most of those found in Tennessee and the eastern U.S. are smaller. Most Texas herds tend to be somewhat larger, probably due to selective breeding for the meat market. In fact, some ranchers have renamed them Tennessee Meat Goats.

They also have a market as pets because they are unique. Myotonic means when they are frightened or excited they "lock up" and often fall over (faint) and lie very stiff for a few seconds. It is an over-simplification, but the chemicals which are rushed to humans' muscles and joints to prepare them for "fight or flight" are withheld in the Myotonic under exciting or frightful circumstances.

No one really knows their origin. There are two theories. One of the possibilities is that a private herd sold to a Tennessee farmer around 1880 was the beginning of the breed. A man named John Tinsley arrived in Marshall County, Tennessee, with four goats, a billy and three nannies, which he had brought from Nova Scotia. When he moved on a year later, he left his goats behind. It is believed that all the Fainting goats in the U.S. can trace their orgins back to these four. The other is that there was a spontaneous mutation of a herd in Tennessee about 1885 which resulted in the recessive gene.

Myotonic goats which have been selected for meat production are heavy rumped, deep chested animals. Most are black and white but multi colors are not uncommon. They breed aseasonally, are easy kidders and have good milk production. Many breeders have noted the breed has the capability to produce two kiddings a year. They are good mothers so in most cases a bonding pen is not needed. Since they are not good climbers and jumpers they are somewhat easier to keep than other goats.

The American Livestock Breed Conservancy has placed this breed of goat on their "rare" list, with an estimated world population of under 10,000. They have now been discovered as excellent crossbreed stock for the Boer goat, a meat goat which was imported from South Africa. The "fainting" gene is recessive, therefore it is usually not expressed in crossbred animals.

Myotonic Goats are also known as Tennessee fainting goats, stiff-legged goats, wooden-legged goats, nervous goats, and epileptic goats.

Myotonia is the condition that causes the Fainting goats to stiffen and/or fall over when they are startled. This has something to do with the absorption and release of calcium in the muscle fibers, and is caused by a combination of recessive and sub-dominant genes. The goats can show different degrees of myotonia. When they are startled, some will go stiff and rigid with their entire bodies, and fall to the ground. They are awake and aware the entire time, and if eating when it happens, will continue chewing their food and swallowing! The condition only affects the external muscles, and lasts for 5-15 seconds, after which they will slowly unstiffen and get up. They usually will walk very stiffly for the first few steps and seem to loosen up as they move, like someone that has been sitting for a long time and whose legs have gone to sleep.

Other goats will only stiffen their hind legs or all 4 legs, but will not fall down. They brace themselves and so remain standing, though stretched out. This seems to be more common among older goats that have learned what to expect, and will even lean against something to keep from falling.

Very young kids often do not show any signs of having it, but as they get older, the condition becomes more pronounced. This is not always the case though, since I have had one faint at 3 days old, and myotonia has been observed while still in the uterus.

The myotonic goat is a medium size breed, although they have been bred by some people down to pygmy size as pets, and other breeders have bred them up to use as meat goats. The myotonia itself expresses in increased muscle mass, and a higher meat-to-bone ratio. Most myotonics, are parasite resistant, and are very efficient feeders---two traits that make them highly desirable as meat producers.

Most will also kid 3 times in 2 years, although extra care must be taken to be sure that a doe is in condition to do this. Twins are the most common, although 3, 4, and even 5 kids at the time have been recorded! Keep in mind that if a goat has over 2 kids, it is sometimes better to bottle feed the extras, or at least supplement them so that the kids all get adequate nutrition. Fainting goats are usually very good mothers, protective and loving with their kids. They kid easily and have a good milk supply. They are a calm and easy-going breed, and they don't jump fences!

The face of a myotonic goat has features that distinguish it from other breeds. The eyes seem to bulge outward more than in other breeds because of the overall structure of the eye socket.

The ears stand out sideways, slightly or wholly facing forward, and sometimes have a ripple about halfway down. This ripple can be prono-unced enough to make the end of the ear drop downward. The goats tend to have a wide muzzle, with a medium length, mostly straight nose. From the side, there is a slight dip at the eye level.

No one knows what the original myotonic goats really looked like, since they are a landrace breed. They are considered a native goat to America, and have developed to grow and breed in the hot, humid Southeast. However, they grow a thick coat of cashmere in the winter, and easily handle winters as far north as Alaska! They come in all colors, with some having long hair, some having short hair, and some having "skirts" of long hair from their top-line and legs.

Some grow short cashmere, and some grow it long enough to shear and spin. It all depends upon what individual breeders like and are breeding for. The only coat that is NOT acceptable is a coat that hangs in long curly ringlets, like an angora.

The origin of fainting goats is still a mystery. They have been traced back to the early 1880's, when a man named John Tinsley showed up in Marshall County, Tennessee, with a sacred cow and 3 nannies and a buck that fainted. By his dress, he was thought to be from Nova Scotia, but never said. He stayed for a while, married, then sold his goats to Dr. H. H. Mayberry and left town with his sacred cow. Dr. Mayberry bred the goats and started the breed in this country. He did a lot of research, writing to veterinarians and foreign ministers of agriculture trying to find out where the goats came from, but had no luck. Recently, new information has come to light, that suggests the original goats that started the breed may have come from somewhere in the mountainous areas of Europe. Research is still being done, since no goats remain there that match the fainter type.

36

Nigerain Dwarf

The Nigerian Dwarf is a miniature goat of West African Origin. It's conformation is similar to that of the larger dairy goat breeds. The parts of the body are in balanced proportion. The nose is straight. The ears are upright. The coat is soft with short to medium hair. Any color or combination of colors is acceptable, though silver agouti (roan) is considered a moderate fault.

Ideal weight is suggested to be about 75 pounds based on HES evaluation information. Animals are disqualified from the show ring for being over-sized for the breed standard, curly coat,roman nose, and pendulous ears or evidence of myatonia.

The most commonly asked question about Dwarf goats is: What is the difference between these and Pygmy goats? Although they have similar origins, they are separate and distinct breeds. Pygmies are bred to be "cobby" and heavy boned. Dwarves are bred to have the length of body and structure, in proportion, of a dairy goat. Dwarf goats come in many colors. Main color families are black, chocolate and gold. Random white markings are common, as are spots and other color combinations such as red, white, gold and black.

Color is one of the big factors that makes breeding Dwarves so popular. You never can be sure what color the babies will be until they are born; even then you can't be sure because many times their color will change.

Black and White Dalmation Patterned Doe

Dwarf goats are gentle and loveable. Even breeding bucks are handled easily. They make wonderful pets and great animal projects for young children in 4-H. Breeders of other types of goats find that their Dwarves blend in with the rest of their herd well and do not need special quarters, just adequate fencing to contain them because of their small size. Dwarf goats breed year round. Many breeders breed their does three times in two years, giving the doe a 6 month plus break. This is of course a personal choice for each breeder. New born kids average about 2 lbs. at birth but grow quickly. They reach sexual maturity at a young age so be sure and separate the bucks and does. Those little guys have been know to breed and be fertile as young as 7 weeks of age.

Does can be bred at 7 to 8 months of age if they have reached good size. Some breeders prefer to wait until they are at lest

1 year or older. Dwarf does can have several kids at a time, 3 and 4 being common and sometime even 5. Dwarfs are generally good mothers and able to take care of their babies should you leave them to do the raising of the kids. They can also provide a surprising amount of milk for their size. They can give three to four pounds per day of 6 to 10% butterfat.

Bucks are able to be used for service as young as 3 months of age and easily by the time they are 7 or 8 months old. Dwarf bucks are vigorous breeders but are gentle enough to be used for hand breeding or pasture breeding. Both methods are used successfully.

37

Oberhasil Goat

The Oberhasli is a breed of dairy goat from the eponymous district of the Canton of Berne (Switzerland). The name loosely translates as 'highlander'. Oberhasli are a standardized color breed, with warm reddish brown accented with a black dorsal stripe, legs, belly, and face. Occasionally a black Oberhasli appears as a result of recessive genes. The American Dairy Goat Association (ADGA) considers black does acceptable and will allow them to be registered, but black bucks cannot be registered.

The breed is about two inches smaller in size than the other standard sized (non-miniature) breeds. Temperament tends to be quiet and sweet natured but alert, with vocalizing similar to the other Swiss-origin breeds. Does can be pushy with other does around feeding time, and in mixed breed herds will likely become herd queen.

This breed is of medium size, vigorous and alert in appearance. Its color is chamois. Does may be black but chamois is preferred. Chamois is described as: Bay - ranging from light to a deep red bay with the later most desirable. A few white hairs through the coat and about the ears are permitted. Markings are to be: two black stripes down the face from above each eye to a black muzzle; forehead nearly all black, black stripes from the base of each ear coming to a point just back of the poll and continuing along the neck and back as a dorsal stripe to the tail; a black belly and udder; black legs below the knees and hocks; ears black inside and bay outside;

bucks often have more black on the head than does, black whiskers, and black hair along the shoulders and lower chest with a mantle of black along the back; bucks frequently have more white hairs through the coat than does.

Oberhasli is still relatively rare in the United States, first appearing as importations in the 1920's, and until the early 1960's, was not yet accepted as a breed by the ADGA. Because of its rarity and close similarity in type to the multicoloured French Alpine goat, the Oberhasli registrations were lumped with the American Alpines. A small number of them were kept pure with records intact, and the breed was kept alive in America almost single-handedly by Esther Oman, a breeder in California.

The Oberhasli breed is growing in popularity at goat shows as their docile nature is appreciated, along with the fine color, described as being 'like the wood on the back of a violin'. Contestants in showmanship classes are discovering that the color is easier to keep clean than pure white animals.

This breed is a good choice for the person who wants dairy goats for milk, from hardy thrifty animals, who appreciates the vivid rust-red coloration, wants something slightly out of the ordinary, and likes the Swiss type head (upright ears).

38

Peacock Goat

Little is known about the origin of the Peacock Goat. The first mention of it was in the year 1887, under the name Prätttigau Goat. This description is for the most part identical to the present outward appearance. Later other names surfaced like gray-black or gray-black-white mountain goat, or razza naz (Tessin) and Colomba (Bergell). The front body half is predominantly white with black boots, the rear half predominantly black.

The inside of the ears and the area around the mouth are dark. On the mostly dark rear portion of the body the white top side of the tail and the white spots on the thighs and flanks are noticeable. The animals wear a thick, mid-length coat of hair and are horned. Characteristic are also the dark facial spots and the dark stripes from the base of the horn over the eyes up to the nose. These stripes or "pfaven" gave the breed its name in that a spelling mistake of a journalist was perceived as fitting for this beautiful animal. During the breed cleansing of 1938 the Peacock Goat was incorrectly considered to be a color variant of the Grisons Striped Goat.

More recent blood testing however could not confirm this assumption. For the near future the Peacock Goat will survive without official recognition as a breed and that without any public funding. For a longer period of time, however, the danger exists that the breeders of the Peacock Goat will increasingly switch to other recognized breeds. For the survival of the breed it is therefore important for the

Federation to rank the Peacock Goat as equal to the other breeds.

This ancient mountain breed was encountered primarily in the canton Graubünden and in Upper Tessin, where it was often put to use as a lead goat in Verzasca herds. Today Peacock Goats are spread throughout all of Switzerland. In milk production the Peacock Goat is equal to most mountain breeds. It produces a good amount of meat with modest fodder demands. In Switzerland about 300 Peacock Goats are kept.

39

Pygmy Goat

The Pygmy Goat was originally called the Cameroon Dwarf Goat. The goat is mostly restricted to the West African countries. Similar forms of Pygmy goats also occur in all of northern Africa, in the south western African countries, and also in east Africa. However, what we call the Cameroon Dwarf goat is the one that we are concerned with and have in the United States. It is the breed that actually came from the former French Cameroon area.

The Cameroon goats were exported from Africa to zoos in Sweden and Germany where they were on display as exotic animals. From there they made their way to England, Canada, and the United States. In 1959, the Rhue family in California and the Catskill Game Farm in New York received the first documented shipments of Pygmy goats from Sweden. Offspring of these animals, as well as earlier imports, were sold to zoos, medical research, and to some private individuals.

Breed Characteristics

A full coat of straight, medium-long hair which varies in density with seasons and climates. On females, beards may be non-existent, sparse, or trimmed. On adult males, abundant hair growth is desirable; the beard should be full, long and flowing, the copious mane draping cape-like across the shoulders. All body colors are acceptable, the predominate coloration is a grizzled (agouti) pattern produced by the intermingling of light and dark hairs, of any color.

Breed-specific markings are required: muzzle, forehead, eyes, and ears are accented in tones lighter than the dark portion of the body in goats of all colors, except goats that are solid black. Front and rear hoofs and cannons are darker than main body coat, as are the crown, dorsal stripe, and martingale; except in goats that are solid black. On all caramel goats, light vertical stripes on front sides of darker socks are required. A pygmy goat is a small breed of domestic goat. Although they produce a large amount of milk for their size, and can be eaten, pygmy goats are not typically used for milk or meat, unlike larger dairy and meat goat breeds. Pygmy goats tend to be more robust and breed more continually throughout the year than either dairy or meat goats. They have stomachs with four compartments: the rumen, the reticulum, the omasum, and the abomasum. As browsers, goats are similar to deer and enjoy variety in their diet, including woody plants.

Females weigh about 23 to 34 kg (35 to 50 lbs) and males about 27 to 39 kg (40 to 60 lb). Their color can range from light/white carmel, medium carmel, dark carmel, dark (red) carmel, silver-light grey agouti, medium grey agouti, dark grey agouti, black with frosted points, solid black, and brown agouti.

Housing

Pet pygmy goats can live in a large dog house or small shed. Their housing should be well-ventilated, but free from drafts. It is convenient to have a concrete floor covered with a rubber mat and sawdust or straw for bedding; the floor should be tilted towards the door to make cleaning simple. In the winter, cleaning is needed less often because goats keep the straw warm as their bed. A bench with dividing slats as suspension will suit the goat as a little bed; if it is very cold, a cheap blanket can be added to add warmth.

Different Coloured Pygmy Goats

A pygmy goat's diet consists of forage, including hay (rough hay is good), and concentrates - i.e., grains such as oats, bran, and barley. It is important to select good quality hay to maximize its nutritional content. Hay, such as grass, alfalfa or oat, should have fine stems and a nice green color. Hay should be fed in feeders that keep it off the ground. Grain or grain mixes supply added protein, vitamins, and minerals.

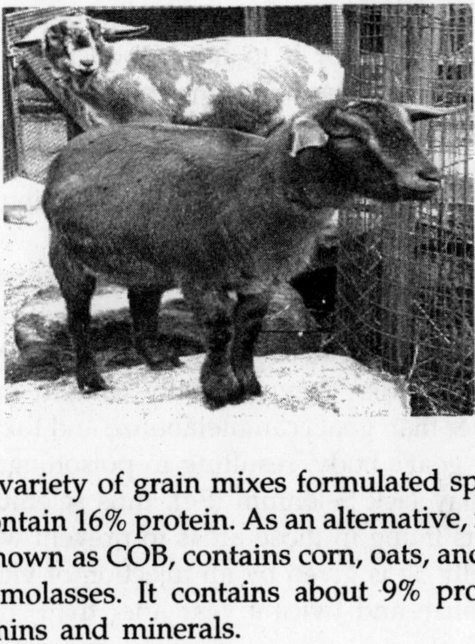

There are a variety of grain mixes formulated specifically for goats that contain 16% protein. As an alternative, rolled ration, otherwise known as COB, contains corn, oats, and barley with or without molasses. It contains about 9% protein and no added vitamins and minerals.

Grain may or may not be necessary as part of a goat's diet, depending on whether one is feeding a buck, wether, or doe. Wethers are susceptible to urinary problems and a low protein grain ration is recommended. Bucks, however, need an adequate amount of grain to maintain sperm production and to keep in good weight. Wethers, once full grown, may do fine with no grain at all, if fed good quality hay.

Female Diet

A doe's needs vary depending on whether she is dry (not

producing milk), pregnant, or lactating (producing milk). Does need adequate protein and nutrients for fetal development and milk production, but too much of a good thing may cause obesity, which can lead to kidding problems and/or a variety of health problems.

Salt and Minerals

Salt and minerals are a vital part of a goat's diet and are found in a variety of sources including hay, pasture, grain, soil, and supplements. Bones are composed of calcium (Ca) and phosphorus (P), therefore, Ca and P must be in a goat's diet in adequate amounts for it to have good bone and muscle development. Even a mature animal must have adequate Ca and P to maintain bone structure. The ratio of Ca and P is quite important-it should be 2:1 in a goat's diet. This ratio is not only important for bone development, but also aids in urinary tract health in male goats.

Salt and minerals are available in block form or loose, and should be available to the goats at all times. Be very careful with feeding supplements to goats that have been developed for horses and cattle. They generally contain much higher levels of copper than goats can metabolize and toxic levels can build up in a goat's body, resulting in poisoning. The soil in some areas may lack Selenium (Se), thus selenium must be added to goats living in those areas to prevent white muscle disease. Usually Se is given by an injection of vitamin E and Se (BoSe) at birth and twice a year after that.

40

Pygoras

Anyone can have a cheerful personality, a sunny disposition and an appealing mischievous bent--but Pygoras don't stop there! Covering their agile bodies are fleecy coats of soft, silky hair. It is this very special fiber that is unique to Pygoras and makes them more than fluffy Chia pets.

Pygoras have inherited the best fiber characteristics of both the Angora goat that produces mohair and the Pygmy that produces a very fine "down". They grow a fleece that can take on many qualities (depending on the fiber type), the fiber maintains its fineness as the animals age, and--when spun--an exquisite yarn results. The fiber is very versatile in that it can be handspun, knitted, crocheted or woven. It also can be felted and/or used to make wigs or beards, etc. on dolls.

Pygora goats produce three distinct types of fleece: type A, similar to the fleece of an Angora goat; type B, more like a blend of the angora fiber and the pygmy fiber; and type C, like cashmere. Between these types, are variations and combinations--just as sheep often produce variations in fleece even within the same breed.

At Verdant Vistas, you will find and may purchase all three fiber types in various stages of processing, i.e., completely unprocessed as it comes from the animal; picked and washed; picked, washed and dehaired; picked, washed, dehaired and blended in rovings; and spun into yarn. Fiber is priced according to type and extent of processing. With advanced notice, you are able to reserve a particular fleece and have it prepared exactly how you want it.

41

Russian White Goat

Dairy goats, of the Saanen and Toggenburg breeds, have been imported periodically from Switzerland and other European countries; these breed undoubtedly contributed much to the formation of various groups and varieties of goats specialized in milk production. In addition, people in both the European and Asian parts of the USSR have been raising milk goats for centuries. Separate groups of goats have formed under the influence of various feeding and management methods, and on the basis of different origins.

Dairy goats in central and northwestern Russia, have strong, dry constitution and a harmonious conformation characteristic of the milk type. They have a light and narrow head. The neck is long and straight and the sacrum wide. The udder is big and smooth; quarters are proportionally developed and teats are wide-set. The legs are strong. The positions of the front and hind limbs is regular. The skeleton is strong. The coat is composed of coarse short, or occasionally long, hair, which is usually white. Animals may be hornless or horned. They resemble Saanen goats in appearance.

Typical of this group are the animals found on private plots in Gorki, Leningrad, Moscow and Yaroslavl regions. They are typical milk goats. Live weight of adult females is 50-60 kg, males 60-75 kg. They average 550 kg of milk per lactation and the best animals up to 1,000 kg. Fat content is high - 4.2-5.3%. Milk goats are prolific: 100 females give birth to 190-220 kids. Some goats kid twice a year. Cases have been recorded of up to six normal kids at one kidding.

Dairy goats in the North Caucasus, Crimea and Central Asia are somewhat smaller, have hair of various colours but are inferior in productivity to the goats of Gorki, Moscow, Yaroslavl and Leningrad regions. There are goats there which have a wool undercoat and, consequently, double utility: they are milked and, in addition, yield about 100-150 g of fiber per head. Milk yields per lactation are in the range 250-400 kg with a fat content of 3.5-5.5%. The average live weight does not exceed 40-42 kg.

42

Saanen

The Saanen dairy goat originated in Switzerland, in the Saanen Valley. Saanen does are heavy milk producers and usually yield 3-4 percent milk fat. It is medium to large in size with rugged bone and plenty of vigor.

Saanen goats are a white breed of goat which are the largest of the dairy breeds. Does typically weigh 150 lb (68 kg) or more, with bucks weighing over 200 lb (91 kg). The Saanen breed also produces the most milk (as a breed- there will be good and poor individuals in any breed) and tends to have a lower butterfat content, about 2.5%-3%.

The Saanen temperament is as a rule, calm and mild mannered; breeders have been know to refer to them as living marshmallows. Saanen goats are easier for children to handle and are popular in the showmanship classes due to their calm nature. With solid white coats and pink skin these extremely mild tempered goats are heavy milkers. They originated from Switzerland.

The Sable Saanen is not crossbred, but is a recessive expression of color derived from the white Saanen. Sables can vary in color from beige through black, with almost any color but pure white. Sables are accepted as a breed in their own right in some dairy goat breed clubs but not in others.

Regardless of color, the Saanen breed is large and big boned but graceful and refined in bone, the ears are erect, and the nose is straight or dished. Both does and bucks usually have beards unless trimmed up for a show, and horns, unless dehorned at birth.

Does should be feminine, and not coarse. Saanens are white or light cream in color, with white preferred. Spots on the skin are not discriminated against. Small spots of color on the hair are allowable, but not desirable. The hair should be short and fine, although a fringe over the spine and thighs is often present. Ears should be erect and alertly carried, preferably pointing forward. The face should be straight or dished. A tendency toward a roman nose is discriminated against. The breed is sensitive to excessive sunlight and performs best in cooler conditions. The provision of shade is essential and tan skin is preferable.

43

Sahelian

The Sahelian belong to the Savanna group of goats with many types and subtypes raised in the Saharan and sub-Saharan region. Thin appearance, narrow body, shallow chest, and sloping short croup. Long spindly legs. Udder as well as scrotum in males split into two halves.

Both sexes horned, most (70%) with wattles, pendent or semi-pendulous ears. Average females weight about 55-75 pounds, height in withers 65-75 cm, heart girth 62-78 cm, males about 85-125 pounds, 65-85cm in withers and 70-85 cm. in heart girth. Stiff short hair, with variety of colors from pure white, cream, to red, black or gray sprinkled or pied, gray, brown or black.

Production Traits

First kidding occurs usually about 18 months, bearing mostly

(70%) one or two kids, with lactation persisting to 5-6 months. Dressing percentage (males and castrates) at 25-30kg liveweight 46-48%. Kept primarily for meat, and skins, milk is only seldom used and never processed.

44

San Clemente Island Goat

San Clemente Island is located off the coast of southern California. It is owned by the U.S. government and used and managed by the U.S. Navy. Feral goats, probably of Spanish orgin, have inhabited the island for several centuries, possibly since the 1500's. Later introductions may have come from the mainland Franciscan missions during the 1600-1700's, while farmers were responsible for later introductions.

The U.S. Navy became responsible for the island in 1934. Hunting and trapping were allowed, but in 1972, when a survey concluded that there were 11,000 goats on the island, a systematic removal program was begun. By 1980 an estimated 4,000 goats still remained on the island.

The Navy then proposed a shooting program to be conducted from helicopters, but was blocked in court by an animal welfare group, the Fund for Animals. This group used helicopters and nets to capture the goats, then took them off the island and found homes for htem across the country. Practically all the goats were removed from the island in this manner.

San Clemente goats are relatively small, close to the maximum standard for dwarf breeds. They are a meat breed, though uncommonly fine-boned and deer-like. They are horned in both sexes. Although the island population once exhibited a wide range of colors and color markings, the goats are now mostly red or tan with black markings.

San Clemente Island goats are a highly endangered breed. Left to run feral on San Clemente Island for a century, they offer genes that are now considered to be an important part of our agricultural heritage. They are small, deer-like, and have kind and gentle dispositions. Their beauty is breathtaking.

San Clemente Island goats make wonderful pets. Just because you take your breeding program seriously, it doesn't mean that you can't have fun with your goats while you're at it. Not very many San Clemente Island goats have been raised as companion animals, so they have a reputation of being shy. This is because the does have passed down their heritage to the kids, and this includes teaching them attitudes. These goats are only shy because of the way they were raised. San Clemente Island goats are by nature sweet-tempered, gentle, and intelligent. A shy or aloof goat will respond well to gentle handling and kindness. A San Clemente Island goat who has had close and positive interaction with people since birth will be as tame and trainable as any other pet goat.

It is recommended that San Clemente Island goats be dam-raised, and never bottle-fed. To tame a baby goat, hold it very, very closely for a few minutes each day in the presence of its mother. Let it become calm with you, and return it to its mother. The kid will bond to you, and this will not interfere with the mother-kid relationship. Although San Clemente

Island goats have not been "bred up" to today's dairy goat standard, there is no reason why they should not have the potential to be good milkers. We're seeking breeders who are interested in milking their San Clemente Island goats to help us survey the breed from a dairy standpoint.

If you're interested in sending in some San Clemente Island goat milk to be tested do the following:

— Milk your goat when the kids have been off of her for a few hours.

— The first two squirts usually go "in the gutter" but you may want to include them for butterfat testing

— Make sure you get a full milking, not just a few squirts. The milk containing the highest butterfat will be at the end of the milking.

— Don't worry if you can't get a lot of milk. Get all you can from your goat, and even if it turns out to be less than 1/4 cup, that should still be enough to send in to the lab.

— Pack a sample of milk and mail it to your local lab (not to us!) including any fee. If your dairy friends start talking about preservers and test tubes, don't let it throw you. Every lab listed on your state's page will also accept samples that are packed in double ziploc bags, or a sterile bottle if you have one Use one of those little blue ice packs to keep it cold in transit, and try to mail it so it arrives earlier in the week, not on a weekend.

— Have the milk tested for fat and solids-Saanens and Alpines run at about 3.5% butterfat, Nigerian Dwarves test at about 6%. The higher the fat, the creamier the milk, the better the cheese. Fresh forage will increase butterfat levels more than grain will.

— And while you're at it, you may want to test the milk for bacteria and have a somatic cell count done.

Then, please contact us with the results. It will help us all get a better profile of San Clemente Island goats. We're posting milk test/state lab contacts on your state's page as we go. Can you milk goats that have supernumerary teats? Sure! One of our breeders milks hers twice daily, and gets about 2½ lbs. per day. You use the same action with your hand, there's just a little more there, and most of it goes where it should.

Another breeder found that two teats on one side of the udder were functional, whereas only one extra teat on the opposite side was functional. Not only that, but she suspects that the side with two functional teats also included two cisterns. Goat meat is delicious. There has always been a demand for goat meat in the United States, and it continues to grow. Goat meat is met with increasing popularity as the country's culinary tastes expand to include more cultural diversity.

Commercial meat goat breeders are also getting interested in San Clemente Island goats. Some of these breeders would like to add the hardiness of San Clemente Island goats into their herds, and these herds may contain percentages of Boer goats, Kiko goats, and others. For those of us who have worked hard to keep purebred San Clemente Island goats going, the thought of allowing them to be purposely mixed with crossbreeds might at first seem like a waste. However, if you do have a good-quality buck that you don't need, that no other breeder needs, and whose genetics are readily available and well-represented, please consider selling him to a commercial breeder who seeks hardiness and who is comfortable with a smaller goat. Spanish goats avoided extinction due to the demand from commercial breeders for their hardy genes, and it would do us well to increase the demand for San Clemente Island goats in similar arenas.

45

Sirohi Goat

This breed is distributed across the Sirohi district of Rajasthan. It also extends to Palanpur in Gujarat. The total goat population in the Sirohi distribution area, according to the 1972 census, was 0.295 m, of which 0.007 m adult males and 0.204 m adult females.

Climate

Average monthly temperature (°C)	Average	Range
minimum	16.5	9.3-22.3
maximum	24.9	19.3-31.5
Average monthly relative humidity (%)		
morning	55	29-95
evening	48	24-91
Annual rainfall (cm)	169	

Breed characteristics

Size

	Adult male	Adult female
Body weight (kg)	50.37 ± 2.52 (16)	22.54 ± 0.17 (343)
Body length (cm)	80.0 ± 1.02 (16)	61.3 ± 0.2 (343)
Height at withers (cm)	85.6 ± 1.4 (16)	68.4 ± 0.2 (343)
Chest girth (cm)	80.3 ± 1.0 (16)	62.4 ± 0.2 (343)

Conformation

Compact, medium-sized animals. Coat colour predominantly brown, with light or dark brown patches; a very few individuals are completely white. Most animals are wattled. Ears are flat and leaf-like, medium-sized and drooping; ear length: 18.8 + 0.6 cm. Both sexes have small horns, curved upward and backward; horn length: 7.7 ± 0.15 cm. Tail is medium in length and curved upward; tail length: 16.7 ± 0.14 cm. Udder is small and round, with small teats placed laterally.

Flock structure

Average flock size is 60, containing 1 adult male, 42 adult females and 17 young.

Reproduction

Under farm conditions, kidding percentage: 89.3%. Litter size: singles: 91.5%; twins: 8.5%.

Breeding

Generally pure breeding. Males are selected on size from within flocks. There is some introduction of Marwari for increasing hair production.

46

Spanish Goat

Goats were a common sight on early European sailing ships. They were desirable because of their relatively small size. They are agile, friendly and capable of eating a much less desirable, and more available, diet than other sources of milk and meat. When Coronado, DeSoto and other Spanish explorers came to America they brought goats as a meat source. Some of these goats either escaped or were released when alternate meat sources were discovered. That would have placed the first goats in Oklahoma and Texas in the early 1540s. Pockets of wild goats, decended from these animals, developed for over 400 years before developing this livestock became profitable.

The trend for smaller, more agile goats was dominant in the wild. Small udders were essential. Large udders would be cut up by cactus, thorns, rocks and any number of other obstacles. These feral goats became known as "Spanish" or "brush goat". Although not of a specific breed ancestry they have developed through natural selection. The term has also become used to describe any goat of unknown ancestry. Most are wild or at least semi-wild. Size varies greatly due to climate, terrain and available breeding stock. Body shape, ear shape, horns, hair and color are not consistant.

Traditionally, in the United States goats have not been considered as a meat source. Ethnic groups have increased in the last few years from areas where goats are a staple in their diet. These recent groups have had the money and influence to attract the owners of wild goats. These owners have begun to change their strategy regarding their goats. Management tactics have been implemented to increase the size and meat characteristics of their animals. New and larger breeding animals, such as the Boer goat, have been imported.

The fat content of the meat is considerably less than beef. America is in need of a healthy red meat source. Producers, consumers and investors are looking at the changes in the animals and the market. The meat goat is capable of becoming the fastest growing livestock animal in quite some time.

47

Tauernsheck

The Tauernsheck is a very rare breed. It's roots are in the Austrian Landrace, the Pinzgauer and an unknown ancestor that brought the white spots. Austrian farmers started breeding in the late1800s. Since 1967 a herdbook has existed. The breed was mostly bred in the Raurisertal (The valley of Rauris) and in the area around the Grossglockner (mountain). In 1994, the Austrian Government could find 20 breeders with around 200 animals. Since then the number has increased. The goat was bred in black and brown, like the Swiss Oberhasli but it has to have a white line in the face and at least two or more big white spots on the sides, that should not break the line on the back. The legs should be brown and black. At least the animals should be bi-color (black and white).

The Tauernsheck was bred in these colors to make it easier to find. Normally every Austrian goat spends the summertime in the Alps. High up in the mountains they stay in their herd and enoy the grass, the good fresh air and the peace. In the fall they are herded back to the farm houses. Goats sometimes try to hide but with triple colors, the Tauernsheck was and still is easy to find up in the Alps. In the summer and fall they are easy to identify by their white spots and in late fall or winter through the black or brown color. The Tauernsheck is a very healthy breed and is known for it's long life.

Both genders have horns and normally do not get dehorned. The does kid once a year with either a single or more usually twins. Tauernshecks are very good mothers and have very high quality milk and the milk production level is very high too. Adult bucks are known for their fighting.

48

Toggenburg

The Toggenburg is a Swiss dairy goat from Toggenburg Valley of Switzerland at Obertoggenburg. They are also credited as being the oldest known dairy goat breed. This breed is medium size, sturdy, vigorous, and alert in appearance. Slightly smaller than the other Alpine breeds, the does weight at least 120lb/55kg.

The hair is short or medium in length, soft, fine, and lying flat. Its color is solid varying from light fawn to dark chocolate with no preference for any shade. Distinct white markings are as follows: white ears with dark spot in middle; two white stripes down the face from above each eye to the muzzle; hind legs white from hocks to hooves; forelegs white from knees downward with a dark lien (band) below knee acceptable; a white triangle on either side of the tail; white spot may be present at root of wattles or in that area if no wattles are present. Varying degrees of cream markings instead of pure white acceptable, but not desirable. The ears are erect and carried forward. Facial lines may be dished or straight, never roman.

Toggenburgs perform best in cooler conditions. They are noted for their excellent udder development and high milk production, and have an average fat test of 3.7 percent.

49

Uzbek Black

Black wool goats in Uzbekistan were produced as a by-product in the formation of the new breed - Soviet Mohair goats. Mating of predominantly white first and second generation crossbred females with purebred white Angora males resulted in the appearance of 1-2% of black kids.

After 4-5 years, there were already several hundred black wool goats; they were gathered into one herd and females were mated only to black males. Such mating produced in the first year 64% and in the second 74% of black kids; others were white, tan, grey, etc. Later, due to positive assortative mating, it was possible to obtain 94% of black kids.

The black goats are similar to Don goats in such aspects as fleece structure, physical properties of wool fibres and productivity. In contrast to Orenburg goats, which have long guard hairs and very short wool fibres, Uzbek goats have wool fibres longer than guard hairs, except along the spine where the latter are very long. Hair is not shed in spring but wool is shed abundantly, starting from the first warm spring days when goats pass to grass feeding.

Wool is combed out selectively at the time of moulting and, as a rule, only once, in mid-March. Average yields per head in collective farms vary from 280 to 440 g. While the length and fineness of fibres depend on the level and quality of feeding, wool yields are actually determined by the time of combing out the moulting fibres. Shedding is very rapid and 5-10 days delay in combing results in 20-40% loss irrespective of age. The extreme limits are 6 and 12 cm in females and even more in males. Fibre diameter is within the range 15-24 μm; the average for females is 19 and for males 22 μm. In the young, wool is 1-2 μm finer than in adults.

The state farm Baisun in Namangan region of Uzbeckistan has a goat herd of 10,000 head and is considered one of the best enterprises for raising black wool goats. Average wool yields are 450 g from females, 700 g from males and 600 g from castrates. Wool goat husbandry in this state farm is very profitable and cost-effective.

50

Valais Blackneck

A truly international breed, the Valais Blackneck Goat is also known in various countries as the Col Noir de Valais, Valaisan a Col Noir, Viege, Chevre des Glaciers, Walliser Schwarzhals, Schwarzweisse Walliser Sattelziege, Vallesana del Collo Nero, Vallese or Valais Blackthroat. Such an international following must surely indicate a highly valued animal!

The Valais Blackneck Goat has traditionally been kept by transhumant pastoralists and is a hardy mountain breed. These goats are efficient grazers who flourish even on poor or scarce ranges. Valais Blacknecks are also extremely hardy, able to tolerate wet or cold climes.

At maturity, a Valais Blackneck buck usually stands 80 centimeters at the withers and weighs around 70 kilograms. Does are shorter at around 75 centimeters and weighing 50 kilograms. They have long, wavy hair and arched, medium length horns. The Valais Blackneck gets its name from its striking coloration. While the goat's body is white from the shoulders back, the head and neck and occasionally the front legs are solid black. The line between the two colors is well defined and impressive. Although Valais Blackneck Goats will produce about 500 kilograms of milk in a 210-day lactation, they are more often used for meat production. The carcass is nicely lean and the muscling is moderately thick. Both the Valais Blackneck's milk and meat productions are extremely impressive given the poor and limited nature of its food intake!

Developed from indigenous goats crossed with Kupferziege goats and improved by selective breeding, the Valais Blackneck has been valued for years. The herd book was established in 1920 and today, apart from the milk and meat qualities of the Valais Blackneck, these goats are valued for their beauty and raised by hobbyists!

Although Valais Blackneck Goats are generally kept in mountain pastures, they should be stabled for the coldest two to six months of the year. While grazing, they usually do not need supplemented feed, but in the winter should be offered grass and fodder. They are agile, surefooted and well adapted to the mountains.

Valais Blackneck Goats reach sexual maturity at an average age and are at their reproductive prime at two and a half years of age. Their litter sizes are average and the kids are hardy, although they have lower rates of gain than do most breeds' young.

51

Zhongwei Goat

Zhongwei goats are produced only in the arid desert steppes of some counties in the Ningxia Hui Autonomous Region and Gansu Province. They live chiefly, if not solely, on salty or sandy plants or shrubs. The kids are slaughtered at 35 days of age for their pelts, which have white, lustrous staples and attractive curls.

Both sexes are horned with the male's horns stretching upward and twisted. The males average 39 kg (85 pounds) and females 24.5 kg (55 pounds). These goats also produce cashmere. The males produce approximately 140 g of cashmere and the female 120 g. The percentage of cashmere in both sexes is 25% of the total fleece. The cashmere length is 7.0 cm and 12.5 microns.

The Zhongwei reach sexual maturity at five to six months and are generally mated at 18 months of age. The kidding percentage is 104-106 percent.

52

Biology of Goat

Anatomy is the branch of biological science that deals with the form and structure of animals. Physiology is the branch that deals with the functions of the body.

Goat owners need to understand the basic structure and functioning of goats if they are to maintain the health and increase the productivity of their herds.

Anatomy of Goat

Points of a Goat

The points of an animal are the salient features that an owner or prospective buyer examines in order to assess its health or its potential as breeding stock.

Skeleton of a Goat

The skeleton is made up of the vertebral column, ribs and skull; the limbs; and the joints. The vertebral column consists of a series of irregularly shaped bones, some fixed and some movable. There are 7 neck (cervical) vertebrae, 13 chest (thoracic) vertebrae, 6 or 7 lumbar vertebrae, 4 pelvic (sacral) vertebrae and between 4 and 8 tail (coccygeal) vertebrae. The ribs are elongated, curved bones that form the ribcage, which is the skeleton of the thoracic cavity (chest). There are 13 pairs of ribs attached to each side of the thoracic vertebrae.

Most of them are fixed-that is, they are joined to the sternum (breastbone) at the front of the chest-but the last two pairs are not attached at the front and are called floating ribs. The skull includes all the bones of the head. It consists of a number of flat bones that form immovable joints, most of which disappear with age; and a lower jawbone (mandible), which forms movable joints with the other parts of the skull. The limbs include forelegs (thoracic limbs) and hind legs (pelvic limbs).

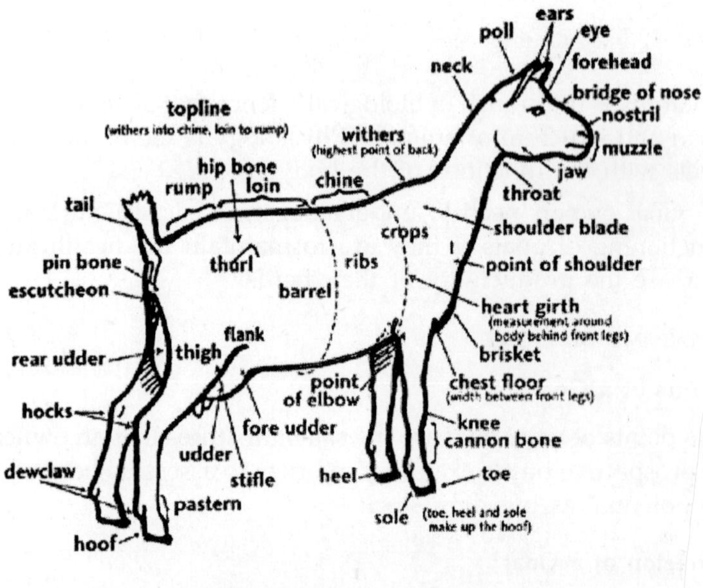

Figure 1. Parts of a dairy goat

The forelegs consist of shoulder, forearm and a lower limb made up of carpus, metacarpus and phalanges. The hind legs consist of pelvic girdle, thigh, and a lower limb made up of tarsus, metatarsus and phalanges. The joints are where two or more bones of cartilages meet. There are two main kinds of joints. Fixed or immovable joints have no joint cavity and are united by fibrous tissue or cartilage.

Many of these are temporary joints which become fixed as calcium accumulates in the flexible tissues and turns to bone (calcifies) with age. Movable joints have a joint cavity surrounded by a joint capsule. There are two opposing bone surfaces, which have smooth surfaces of very dense bone. Movable joints have cartilages between the bones to reduce friction and absorb concussion.

The inner lining of the joint capsule is the synovial layer, which produces synovial fluid ('joint fluid'). Examples include joints of the limbs, such as the shoulder and the hip, and the joint between the skull and the vertebral column. There are also joints with limited movement and rigid supporting tissues; for example, the joints between the vertebrae in the spinal column.

Digestive System of Goat

Teeth

Goats have three groups of teeth:

— four pairs of incisors on the lower jaw only, which bite on to a 'dental pad' on the upper jaw;

— three premolars on each side of the upper and lower jaws;

— three molars on each side of the upper and lower jaws.

The teeth usually erupt at about the times shown below.

	Temporary	Permanent
First incisors	1st week	13-15 months
Second incisors	2nd week	18-21 months
Third incisors	3rd week	22-24 months
Fourth incisors	4th week	27-32 months
Premolars	2-6 weeks	18-24 months
First molars		3-5 months
Second molars		9-12 months
Third molars		18-24 months

A goat's age can be estimated by examining the permanent teeth that have erupted.

Stomach

The stomach of a goat is very large and consists of four parts: the rumen, the reticulum, the omasum and the abomasum. The rumen occupies almost all of the left half of the abdominal cavity. It is divided internally into compartments by pillars (large folds of the wall) and lined with many papillae. The rumen wall has a layer of muscle.

The reticulum is the foremost part of the stomach. It lies beneath ribs 6-8, just behind the diaphragm and liver. Its lining is arranged in a mesh-like 'honeycomb' pattern. The omasum is the smallest part of the stomach. It is oval in shape and sits on the right side of the abdominal cavity. Its interior is divided by large longitudinal folds of tissue, like pages in a book. The abomasum is about twice the size of the reticulum, but elongated in shape. It lies on the right side of the abdominal cavity. The lower end of the abomasum is constricted by the pyloric sphincter, a strong muscular valve which separates it from the duodenum.

Intestines

The small intestine is about 25m long, with an average diameter of 25mm. The first 60cm forms the duodenum, which is next to the pancreas. The small intestine lies at the back of the right side of the abdominal cavity. It is made up of the jejunum and the ileum.

The caecum is a blind sac which marks the junction between the small intestine (ileum) and the start of the large intestine. It is about 200mm long and 50mm in diameter. The colon, or large intestine, is about 5m long. At its beginning it is the same diameter as the caecum, gradually tapering to about the same as the small intestine, when it becomes the rectum.

Reproductive System-Female

The external part of the female genitalia is called the vulva. It seals the opening of the reproductive system to protect against contamination and to prevent the sensitive internal tissues from drying out. It is pinkish grey in colour. The vulva leads to the vagina, which is a tubular structure with a smooth, moist, glistening lining. The lining is tough and covered with mucus to withstand possible injury and bacterial contamination resulting from natural service. The cervix is a series of muscular ridges forming a protective block between the uterus and the exterior. To keep out harmful bacteria, foreign debris and undesirable fluids there is a cervical plug of mucus-a thick, tacky secretion from the mucus glands in the lining of the cervix. This plug seals the cervix during pregnancy and when the doe is not on heat.

The uterus is often called the womb. When the doe is not pregnant, the uterus is 2-3 cm long. It has two horns extending from it. The lining of the uterus is dark pink because it is covered with tiny blood vessels that are there to provide nutrients for the growing embryo. The caruncles-small raised areas on the uterine wall-are the points of attachment for the placenta.

The two uterine horns are linked by a thin, web-like ligament until they curve downward and backward, when they separate and taper off into the fallopian tubes. These lead to the ovaries, to which they are attached by funnel-shaped structures called infundibula. The infundibula and the fallopian tubes are lined with tiny, hair-like projections called cilia, which wave in unison to create currents in the flow of mucus. The sperm find the egg by swimming directly against this current.

Reproductive System-Male

The scrotum is the sac of skin which contains the two testicles (testes). The testes are ovoid in shape. They grow to about 10cm long and usually weigh 250-300 grams. The epididymis runs from the top of the testicle down the side to the bottom (or tail), then upwards again through the inguinal canal to become the vas deference. The epididymis and the vas deference transport semen. The penis has a sigmoid (S-shaped) flexure which allows it to lengthen during mating. The urethra extends 2-3 cm beyond the end of the penis. The accessory sex organs-the seminal vesicles, the prostate gland and the ampullae-are attached to the vas deference and lie next to the bladder.

Structures of the Skin

The follicles are the fibre-bearing structures of the skin. They are formed in the foetus during pregnancy. The primary follicles have hair fibres at birth, and the secondary follicles start to produce fibre during the first few months of the goat's life. The primary follicles secrete sweat from the sudoriferous glands and wax from the sebaceous glands.

The S/P ratio is the ratio of secondary follicles to primary follicles. Each primary follicle has an associated group of secondary follicles. The density of the follicles is the number of follicles (primary and secondary) per square millimetre of skin. Density indicates the potential fleece production.

Feet

The wall of the hoof is horn tissue produced by the horn bud cells around the coronary band (the junction of the skin and the wall of the hoof). The horn grows downward as the wall wears away during walking. The weight of the goat falls on the wall of the hoof.

Embryology

The zygote is a fertilised egg (ovum). The morula is a 16 or 32 cell structure formed by the division and redivision of the zygote. The bastocyst is a structure with a fluid-filled cavity (blastocoele) surrounded by the cells formed into a layer. The inner cell mass is the layer of cells pushed to the side wall by the fluid; this is the mound of cells that will form the embryo. The cells differentiate to form the specific organs in the embryo. The first stage of differentiation is three layers of primitive cells:

— The ectoderm (outer layer), from which form the nervous system (including the brain), sense organs, skin, hooves, hair and mammary glands;

— The mesoderm (middle layer), from which form the circulatory system (including the heart), skeleton, muscles, kidneys and the reproductive tract;

— The endoderm (inner layer), from which form the digestive system, liver and lungs.

The amnion and the allantochorion are membranes around the outside of the embryo. The amnion contains fluids in which the embryo is suspended. The allantois-a sac which accumulates waste products in fluid-combines with the chorion (the outer layer of membrane) to form the allantochorion. The allantochorion attaches to the lining of the uterus to form the placenta. The cotyledons are the points of attachment of the placenta to the uterine wall. They include the caruncles, which are extensions of the uterine wall.

After differentiation of the tissues and organs, the embryo becomes known as a foetus. The circulatory system develops rapidly, and by the third week an embryonic heartbeat is detectable. Dizygous (non-identical) twins form from the fertilisation of two ovulations in the same oestrus cycle; monozygous twins result from the fertilisation of a single ovum and are genetically identical.

Goat's Udder

The goat's udder is in the inguinal region (the groin). It has two glands, one on either side of the midline. Each gland has a single teat and each teat has a single orifice (opening). The glands are separate units. The milk-producing cells are grouped into lobes, and within these are lobules. These are further divided into alveoli, which are tiny sac-like structures. They have an internal cavity called a lumen, and are lined by milk-secreting epithelial cells. Each alveolus is surrounded by muscle cells which contract during 'milk letdown'.

PHYSIOLOGY

Liver Functions

— To excrete bile, which passes into the gall bladder and thence into the small intestine to aid digestion.

— To store glycogen, the source of energy.
— To store minerals such as copper and iron.
— To break down aged red blood cells.
— To break down toxic substances.
— To break down waste products from body metabolism.

Kidney Functions

— To filter out and excrete waste and excess materials from the bloodstream.
— To maintain the fluid status of the body by removing or retaining fluid, and to maintain the correct pH level and chemical balance.

Reproduction

Goats mature sexually at 5-8 months of age. Reproduction depends on successful deposition of the buck's semen in the vagina of the doe at the correct time in her oestrus cycle. Every 18-25 days a fluid-filled follicle containing oestrogen develops in the ovary. The oestrogen is carried throughout the body by the bloodstream. It brings the doe on heat by affecting certain tissues that play a part in fertilisation: for example, the glands lining the cervix, which produce a clear stringy mucus and an attractive (to the buck) odour. Also, oestrogen stimulates in the brain of the doe a desire to accept the buck.

Heat usually lasts for 24-36 hours. Ovulation usually occurs towards the end of heat. The egg is released from the follicle to enter the fallopian tube, with the assistance of the funnel-shaped infundibulum. For successful fertilisation, the sperm must travel from the cervix through the uterus and along the uterine horn to the fallopian tube into which the egg is extruded. If the heat periods are not regular, ovulation is unlikely to coincide with them and conception becomes difficult.

The function of the ovary is to produce an egg every 20 days (normal range 19-21 days), and the regular onset of heat to prepare the doe at this interval is essential. One part of this preparation is the action of oestrogen on the muscle fibres in the uterus, making them sensitive to the hormone oxytocin, which in turn makes the muscles contract in a wave-like action. This helps the sperm to reach the egg.

This assistance for the sperm is essential, since its average life span in the female genital tract is about 24 hours. The sperm travels against the flow of the uterine mucus, and would certainly perish without reaching the fallopian tubes if its only method of propulsion was the wriggle of its tail. So the wavelike action moves the sperm up the genital tract against the flow of mucus and towards the ovary. Once released, the egg perishes within 12 hours if it is not fertilised.

After ovulation the cells in the ovary produce a different hormone, progesterone. This takes the doe out of heat, stops the wave-like muscle contractions and helps to prepare the uterine wall for implantation of the fertilised zygote. Progesterone is produced in the cavity left by the ruptured follicle, which fills with cells to form the yellow body. This continues to secrete progesterone for 13-14 days. If the egg is fertilised, the corpus luteum goes on to play a part in early gestation. However, if fertilisation does not take place the uterine wall secretes prostaglandin, a substance which causes the corpus luteum to regress. This occurs between day 16 and day 18, thus completing a full cycle. A new follicle begins to develop and produce oestrogen over these days to bring the doe into heat again at about day twenty.

Digestion Process

Goats are ruminants. They eat plants, including pastures (grazing) and many weeds, shrubs and trees (browsing). Digestion is the process of breaking down this plant material in the stomach and intestines into components which can be absorbed and used by the goat.

The rumen works by:

— muscular action-contractions of the muscular walls of the rumen break down newly ingested grass and mix it with rumen liquor; and

- — microbial action-bacteria and protozoa secrete enzymes which set up a chemical breakdown process commonly referred to as fermentation. The walls of the rumen secrete no digestive juices. Any changes to ingested material which occur in the rumen are the result of digestion by the microbes that live in the animal's gut.
- — Carbohydrates are converted into fatty acids such as acetic, propionic and butyric acids. These have simple chemical structures and are known as volatile fatty acids. Many of these acid molecules are absorbed directly into the blood through the walls of the rumen.
- — Proteins are broken down into their constituent parts, called amino acids. These amino acids are further broken down into fatty acids and ammonia, which can be used by microorganisms to build microbial protein. This protein may contain some amino acids which were not present in the dietary protein.
- — Non-protein nitrogen, such as urea, is broken down in the rumen to yield ammonia. This is later absorbed by microorganisms to make microbial protein.

The fermentation processes in the rumen produce large amounts of gas, mainly in the forms of carbon dioxide and methane. These are normally expelled by belching. Strong muscular contractions of the rumen cause the rumen liquor to spill into the reticulum in small amounts at a time. This liquor contains digested food particles which pass through the reticulum and into the omasum. The omasum has leaf-like partitions which line its walls and absorb water from the ingested material which passes into it. The abomasum is the ruminant's only true stomach. It secretes gastric juice containing hydrochloric acid, which kills the rumen microbes and begins to break them down. The gastric juice also contains pepsin, which starts the digestion of proteins. In the small intestine, enzymes from the pancreas and the walls of the small intestine act on ingested matter and the microbial cells contained in it. The nutrients are then absorbed. It has often been said that ruminants live not on pasture but on dead rumen microbes which have predigested the pasture.

53

Dairy Goat Production

Dairy goat production is an alternative livestock enterprise suitable for many small-scale or part-time livestock operations. Some dairy goat producers have been successful in pasteurising goat milk and building an on-farm jugging business, while others have ventured into processed milk products for retail distribution, especially speciality cheeses and yogurt. The potential also exists for selling milk to processors, usually on a regional basis. Although fluid milk and processed products are important markets, dairy goat producers should also consider the potential for selling animals to hobbyists and youth involved in vocational agriculture livestock projects.

It is quite certain that the goat was one of the earliest domesticated animals in Western Asia. The goat is thought to have descended from the Pasang or Grecian Ibex, a species of wild goat found in Asia Minor, Persia and other nearby countries. A reference to the use of mohair from goats can be found in the Bible at the time of Moses when he told the children of Israel to bring white silk and goat's wool to weave altar cloths for the Tabernacle. The first goat importations into this country came from Switzerland. Records of early settlements in Virginia and New England indicate that milk goats were brought to the United States by Captain John Smith and by Lord Delaware.

There were very few Swiss goats in the United States prior to 1904. In that year, a consignment of ten Saanens and 16

Toggenburg goats were brought in, followed by more in the next two decades. These goats spread all over the country and have provided the basis for the development and improvement of milk goats in this country. In the United States, milk goats are found in every state, with California, Wisconsin, Texas and several eastern states leading in goat population. In recent years, Florida has been rapidly becoming a popular state for dairy goat breeders.

The American Dairy Goat Association recognises six breeds of dairy goats in the United States. They are Alpine, American LaMancha, Nubian, Saanan and Toggenburg. A lesser known dairy breed is the Oberhasli. They range in color from light to deep red with black stripes down the forehead and black legs below the knees.

The Alpines are composed of several varieties including the most popular French Alpine and the less numerous British, Rock and Swiss Alpines. They are medium to large in size, with color variations from pure white through shades of fawn, gray, brown, black, red buff, and combinations of these colours in the same animal.

The LaMancha was developed in the U.S. from crossing a short-eared goat of Spanish origin with purebreds of the Swiss breeds.

Figure 1. LaMancha

They are distinguished by very short, or the total lack of, external ears. There are two types of La Mancha ears: the "gopher" ear and the "elf" ear.

The allowable length of the elf ear is about two inches and the gopher ear is one inch. There is no preference for ear type in does, but bucks must have the "gopher" ear type to be eligible for registration. The American LaMancha is medium in size with any color or combination of colors and a short, glossy hair coat. (Figure 1)

The Nubian is a relatively large, proud, and graceful dairy goat of Oriental origin, known for high quality, high milk fat percentage. They are characterised by long drooping ears, convex noses and a short, glossy hair coat. Since any color or pattern is allowed, it is the distinctive convex facial profile between the eyes and the muzzle, and the long bell-shaped ears, that immediately identify the breed.

Figure 2. Nubian

The ears should extend about one-inch beyond the muzzle. They are medium to moderately large in size with good stature. (Figure 2)

The Saanens are white or light cream in color with white preferred. Spots on the skin are not discriminated against. Small spots of color on the hair are allowable, but not desirable. They are medium to large in size with rugged bone, plenty of vigor yet feminine throughout. The ears should be of medium size and carried erect. A tendency toward a roman nose is discriminated against. (Figure 3)

Figure 3. Saanens

The Toggenburgs are medium in size, sturdy and vigorous. The Toggenburg may be various shades of brown with white markings.

Figure 4. Toggenburgs

These white markings appear as two stripes on the face from eye area to nose, around the outer edge of the ear, from foot to knee, from foot to hock and as a triangular patch on each side of the tail, covering the pin bone area. Ears are of medium size and alertly carried. (Figure 4)

The Oberhasli dairy goat breed is a newer breed to Florida. The dairy goats are of medium size, vigorous and alert in appearance. The color preferred is chamois but does may be black. The chamois color varies from a light bay to a deep red bay with the latter most desirable.

Figure 5. Oberhasli

Markings include two black stripes down the face with the forehead nearly all black, black legs below the knees and hocks, and ears black inside and bay outside. The face is straight. (Figure 5)

Commercial Dairy Goat Operations

The greatest number of goat owners in this country keep goats for family purposes. In Florida, there are three licensed commercial dairy goat operations of which two are processing and marketing Grade A pasteurised goat milk and the other is marketing speciality products such as cheese. Most of the operations in Florida are rather small with less than ten milking does. Many of the dairy goats are registered and are exhibited at various shows throughout the state. The more popular breeds are Nubian, Alpine and LaMancha. The Florida Dairy Goat Association and the annual Florida Dairy Goat Production Conference have been helpful in recent years in providing needed materials and information to dairy goat producers.

The selection of milk goats for commercial dairy herds and breeding herds must be accomplished in terms of keeping and maintaining good records, and becoming familiar with the different strengths that dairy goats possess. They may or may not be registered, but must be able to produce milk over a long period of time with good care. Good herds frequently consist of goats having herd averages of 1,500 to 3,000 pounds of milk. The top three dairy goat herds in Florida in 1992 completed DHI 305 day lactation records above 2,100 pounds of milk with the top herd at 2,934 pounds.

For those that are interested in getting involved in raising or showing dairy goats, start with two kids (three to five weeks old) so they can get to know you and you can have the enjoyment of watching them grow. It is best that they be of the same sex, preferably does if you plan to expand. If a buck is preferred, have the breeder castrate and dehorn the animal prior to taking it home. Also at this young age, about a quart of milk is needed daily until about 8 weeks of age. At the time of weaning, the doeling should be eating some grain on a daily basis. The choice of breed is purely a personal one.

Artificial Insemination

The normal breeding season for the dairy goat is from late August to mid-March. Most of the does are bred in September, October and November and produce offspring in February, March and April. Because goats are seasonal breeders, the milk supply may be short for 2-3 months during the late fall and winter months. Planned matings through good record keeping and a sound breeding programme will largely overcome this.

Time of Breeding

Doelings reach puberty by 6 to 8 months of age and are usually bred at 7 to 10 months of age. At the time of breeding they should weigh about 80-90 pounds (60% of adult weight). If the doelings are not at an optimum weight, breeding should be delayed since puberty is more dependant on body size than age. However, delaying breeding much after 10 months of age decreases the reproductive performance. Growth rates of replacements should be monitored and their nutrition adjusted accordingly. For two to three weeks prior to the breeding season does and doelings should be gaining weight. This is achieved by increasing the amount of energy being fed. Does managed in this manner will have an increased number of ovulations.

Signs of Estrus

Estrus is the period when the doe will receive the buck (or the time of Artificial Insemination). Usually this period will last from a few hours to 2-3 days and is characterised by frequent and insistent "talking", tail wagging, and pink color as well as swelling in the external genital region-sometimes with a discharge. A lactating doe will usually drop in her milk production. The period between estrus' is from 17-21 days. To achieve the highest conception rate, it is best to breed the doe on day two of her estrus period.

During the breeding season bucks have a strong odour and should be kept in separate pens at all times. This will aid you in being able to get the does bred over an extended period of time in order that milk will be available over a 12-month period and reduce the possibility of off flavours in the milk. Maintain good records and record all heat periods. Breed to the best buck available if you plan to retain the kids. Any notes of her behaviour may help in out-of-season breeding or in detecting estrus early in the next heat period.

Gestation Period

The gestation period is the time from conception to kidding. Normally this is a period of 145-155 days or on the average 5 months. Does producing milk at the time of breeding will be at their peak production. The doe is usually bred at 45 to 60 days in lactation and will normally be producing milk over a 305 day period. The kidding interval should be about 12 months.

Artificial Insemination (AI)

Artificial Insemination (AI) is nothing more than the placement of semen in the female reproductive tract by mechanical means rather than by natural mating. While the term AI does not necessarily imply that the semen used be preserved through prior freezing, this is the most common situation in dairy goats. The main value of AI lies in its use as a tool for the improvement of livestock, by extending the use of bucks that possess desirable characteristics. There is probably no other sufficient reason to consider AI.

Dairy Goat Kids

Kids are born without antibodies circulating in their blood and rely on antibodies in colostrum, or first milk, for protection against disease during the first few weeks of life. The antibodies are concentrated in the doe's udder prior to kidding, are sucked by the kid and then passed through the intestinal wall into the kid's circulation. This transfer, or absorption, is made possible by special cells in the intestinal lining that permit antibodies to pass through for the first 18 hours of life. After that time, the cells are eliminated and no further antibodies can enter the blood stream. It is critical then, that kids suckle soon after birth, preferably within 2-4 hours. Colostrum is also high in nutrient value, especially vitamin A, B-vitamins, proteins, and minerals.

The protein content of colostrum is about 20% as compared to 3.5% for normal milk. Overfeeding colostrum or other milk to kids can cause loose bowels and possibly scours. The extra colostrum should be placed in the refrigerator and fed later at about body temperature. The kid must be handled gently and not forced to drink. After a few hours, the hungry kid will drink readily. The kid may be changed to goat's milk, cow's milk, or powdered milk after about one day on colostrum. Provide about 2 to 3 pints of milk each day in 3 to 4 feedings the first 2 to 3 days and twice per day thereafter. At birth, the kid weighs approximately 7 to 9 pounds, heart girth 14 to 15 inches, and height at withers 14 to 15 inches. The kid must be treated as a simple stomach animal such as the dog or cat. That is, a milk diet is needed for the first few weeks of life.

A small amount of grain such as a calf starter or goat chow may be introduced to the kid at 2 to 3 weeks of age. In general, the grain should contain about 14 to 15% crude protein with added minerals and vitamins. As soon as the kid starts eating, the rumen starts developing and eventually the kid will start chewing its cud. This is an indication that all four compartments of the stomach (rumen, reticulum, omasum and abomasum) are developing. Animals having four-compartment stomachs are referred to as ruminants. As the animal grows, the rumen becomes the largest compartment.

Clean, fresh water and salt blocks should be available at all times and especially as the kid is weaned from receiving milk at 8 to 12 weeks of age. Start the kids drinking from a bucket as you discontinue milk feeding. Also, be sure the kid has started eating some grain and hay. All kids should be dehorned at 2 days to 1 week of age, except those that are naturally hornless. Electric dehorners are frequently used with excellent success. Instructions may be received from your County Extension Agent or a local dairyman. The process appears painful but is rather short and causes no harm. An ointment may be applied to the burned area. Dehorning may also be done by a veterinarian under anaesthesia. This allows for a more accurate operation with no pain to the kid.

Dairy Goat Management

Goats are energetic, inquisitive and versatile in their feeding habits. An area near the barn that provides some browse materials (trees, leaves, bushes, twigs, etc.) appears to be advantageous and enjoyed by dairy goats. The importance of such materials toward the nutritional requirements of lactating dairy goats is probably quite small, especially where a fairly large number of dairy goats are being maintained. Good quality hay and a balanced grain mix appear to be the best approach in maintaining high levels of milk production. Fiber in the total ration is needed to maintain a normal milkfat test. However, too much poor quality fiber will lead to lowered levels of milk production. Rations containing some cottonseed hulls or other fibers may be included in the grain where hay or other roughages are not readily available.

The success of a dairy goat enterprise is dependent on the establishment of a good feeding and management programme. Nonlactating dairy goats do well on good pastures supplemented with minerals. Permanent pastures may consist of several varieties of grasses including bermuda, bahia and pangolagrass and a number of legumes including clovers and perennial peanuts. Summer annual grasses include several varieties of Pearl millet and sorghum-sudangrass. Both are erect, tall-growing, high-producing annuals. Sorghum-sudangrass is adapted to drier soils than millet.

Winter annual grasses include rye, oats, wheat, barley and ryegrass. Excellent grazing may be obtained from winter grasses under conditions suitable for growing them. Legumes make excellent hay and are usually superior to most other hay crops because of their higher protein content. While alfalfa hay is the best legume hay, variations may occur in quality. Good quality alfalfa hay should have a green color, small stem, adequate in leaves and 17-20% protein. Other legume hays that may be available are alyce clover and perennial peanut.

Grass hay usually varies considerably in quality. Even so, some grass hay may be used in the ration with success. Grass hay as the only roughage source may limit maximum performance. In addition to pasture and/or the forages being fed, the overall ration should be balanced with a good grain concentrate that is fortified with minerals and vitamins. The exact amount needed will vary with pasture and forage quality. Dairy goats are good eaters and can consume from 4 to 7% dry matter (DM) per 100 lbs body weight as compared to 3-4% DM consumption for dairy cows. This high level of intake allows the dairy goat to have an abundance of nutrients readily available for the synthesis of milk. Overall, the efficiency of milk production by the dairy goat is quite similar to that of the dairy cow.

Important factors to consider when selecting grain rations for dairy goats is quality of forage or roughage being fed as well as the kind of forage. With most grasses and silages in Florida, an 18-20% protein grain mixture is needed to supplement the forage since corn and/or sorghum silage and average quality grass hays are low in digestible protein. With good quality hay (15-20% protein) or the use of small grain pastures such as oat, rye and ryegrass, a 14-16% protein grain mix is generally adequate. Both dairy goats and dairy cows are ruminant animals and can therefore eat the same kind of ration. Common ingredients used are cracked corn, rolled oats, cane molasses, cottonseed hulls, soybean meal, wheat midds and soybean hulls. Assistance may be needed in balancing and formulating the overall ration to prevent problems associated with poor nutrition.

Dry Period

The doe should be bred to freshen once each year with a dry period of about two months. The dry period allows the mammary system time to repair and regenerate for the next lactation. The greater her production the more likely that her body has been depleted of the nutrients used in milk secretion and the longer the dry period required to replenish the losses and store adequate reserves for the next lactation. Does which are not given a normal dry period usually produce only 65 to 75% as much milk in the subsequent lactation as does given a dry period. Grain consumption should be reduced or removed near the time that the dairy goat is turned dry.

At the time of drying-off, substitute fair to good quality grass hay for alfalfa or other sources of legume hay. The dairy goat should be down to about three pounds of milk per day or less. Her milk flow will be reduced quicker if you change her routine at the same time you discontinue milking. Since mastitis is frequently started during the dry period, check her udder for possible problems such as mastitis. A certain amount of pressure is needed in the udder in order to stop milk secretion and flow. If the doe has had mastitis during her lactation, this is the best time to treat it with an appropriate antibiotic for treating dry cows or goats. The dairy goat should be maintained in good condition during the dry period so she will freshen in a healthy state and have every opportunity to produce more milk in the next lactation.

A good mineral mixture should be available. A purchased mineral may contain from 12 to 18% calcium, 6 to 8% phosphorus and 25 to 30% salt, with trace minerals and vitamins. The exact ratio of calcium to phosphorus needed is dependent on forage source since legumes are higher in calcium than grasses. During the last 3 to 4 weeks of gestation, nutrition becomes more important to the doe. She should receive a better quality grass hay and about the same type of ration she will receive after kidding. The doe should be managed during the dry period so that she is in good condition at the time of kidding. She should not be allowed to become fat. The key to success is to have the doe kid in a healthy condition and with a fairly good appetite.

Types of Housing

A small barn or shed is needed to reduce the exposure of goats to wind, rain and solar radiation. The type of housing needed will vary with the number of goats owned and the convenience preferred. A stanchion or box stall with built-in feeder may be advantageous for milking. Does are usually milked on a stand 12 to 18 inches high or higher if preferred. The stand should be long enough for the doe to stand comfortably and about 18 to 20 inches in width. The stanchion should be placed at the front of the stand in order to fasten the doe's head. A small runway may be constructed to the stand to reduce possible udder injury. It is best to construct in an area that can be cleaned easily.

Internal and External Parasites

Internal Parasites: It is important that your dairy goats live in a good and clean environment. A good parasite control programme should be initiated as soon as you obtain your first animal. The reason is because parasites begin to grow in numbers and will build-up on closely eaten grassy areas near the barn. Pastures and lots remain contaminated for long periods and goats may pick up the parasites. Inside the animal, they can interfere with nutrients, cause diarrhea, or result in poor performance of the animal. A good worming programme is a necessity for successful dairy goat farming in Florida.

External Parasites: Parasites including lice, ticks, horn flies, house flies, stable flies, horse flies, deer flies and mosquitoes present serious problems during the year. These pests are most prevalent during the spring and summer months. Many are a problem throughout the entire year in Florida. Lice, both biting and sucking varieties, may present a real problem if not controlled. Since insecticides are frequently changed and taken off the market, contact your local County Extension Agent about current recommendations.

Composition of Goat's Milk

The composition of goat's milk varies both within and between breeds. Various values have been reported for each of the nutrients. This has undoubtedly resulted from analysing milk from a single breed, a single herd, or the analytical techniques used. Goat's milk contains more fat and ash than cow's milk, but has less lactose. Generally, the composition of goat's milk can be expected to fall within a specified range for each milk component. Fat, the most variable component, will usually fall between 3.0 to 6.0% in herd samples. However, values outside this range are not uncommon for individual samples. The ranges that can be expected for total solids, protein, lactose, and ash are 12-16, 3-4, 3.8-4.8 and 0.70-0.95 respectively.

Protein in Goat's Milk

The protein in goat's milk can be divided into casein and whey protein. Casein accounts for about 83% of the total protein and is the primary protein fraction in cheese products. Casein will coagulate under certain conditions and can be removed from the milk. Rennet, acid, and a combination of pepsin and acid will all coagulate casein in milk. Each method closely resembles a natural process of casein coagulation. Rennet coagulation is the process that is used in cheese making. The addition of acid increases the acidity of milk until the casein coagulates in the same manner as sour milk, and the human digestive process is stimulated by the acid-pepsin coagulation of casein.

Whey is the clear liquid that remains after casein is removed from the milk. Proteins that remain in the whey are the whey proteins. Both casein and whey proteins are general categories of proteins. Each contain many individual proteins. Many of these proteins are similar to cow proteins and cause identical allergic reactions. However, there are specific proteins in goat's milk and these are immunologically distinct from proteins in cow's milk.

Curd formed in goat's milk with acid-pepsin treatment is apparently softer than the similarly formed curd of cow's milk. However, curd formed with rennet appears to be stronger in goat's milk than in cow's milk. Curd strength varies between individual animals and lactation. Curd strength decreases to minimum in mid-lactation and then increases to the end of lactation. Although the average percentage of fat in goat's milk is 4.25% it varies with individual animals, breeds, state of lactation, and type of feed. Goat's milk fat contains appreciable amounts of caproic, caprylic, and capric fatty acids.

Although these fatty acids are not unique to the goat, they are more abundant in goat's milk than milk from other species. They are responsible for the characteristic flavour and odour of cheeses made from goat's milk. Goat's milk contains a higher proportion of small fat globules than cow's milk but is similar to sheep's milk in this respect. This has been interpreted as the reason for the slow creaming of goat's milk. However, the primary reason for slow creaming is the absence of fat globule clustering. Cow's milk contains a protein, not found in goat's milk, that causes fat globules to cluster, thus creaming at a rapid rate.

Vitamins in the Milk

Considerable information has been compiled concerning the vitamins in the milk of various species. The primary difference between goat's milk and cow's milk is the much lower concentration of vitamins B 6 and B 12 in goat's milk. However, when considering the use of goat's milk for infant food it bears consideration that goat's milk is nearly as high in vitamin B 6 and twice as high in vitamin B 12 as human milk. Cow's milk is extremely low in vitamin D and none. However, most commercial milk is fortified with vitamin D. It is very interesting to note that vitamin A in goat's milk exists exclusively as vitamin A and not carotenoid pigments. Carotenoid pigments are precursors of vitamin A and are present at varying levels in cow's milk depending upon the breed. Carotenoid pigments cause fat to have various degrees of yellow colouring. Their absence in goat's milk causes butter made from goat's cream to be white.

Production of High Quality Milk

All hair should be clipped closely on the udder and flank area for cleanliness and health reasons. Also, milk taken from such animals is easier to keep clean. Herds should be checked for brucellosis and tuberculosis to insure that these potential human pathogens are not present. If these diseases are discovered the animals should be removed from the herd.

The milking procedures should follow recommended sanitation practices whether it is hand or machine milking. The udder and particularly the teats should be washed with warm water (110 degrees F) that contains an appropriate sanitiser. This not only stimulates the animal for milking, but also has the potential of destroying organisms on the teat that might contaminate the milk. The first few strips of milk from each teat should be examined for abnormalities with a strip plate. Milking, whether by machine or hand, should begin within 2-3 minutes of washing the udder. If milking is done by hand, special precautions should be taken to prevent contamination of the milk. A hooded pail is an asset for hand milking. After milking it is desirable to dip each teat in a dairy teat dip. Teat dips have proven quite successful in the prevention and reduction of mastitis.

Milk should be filtered through commercial filters and cooled immediately. If commercial milk cooling equipment is available, the milk should be cooled to 35 degrees F as quickly as possible. However, if this is not available, milk will cool much more rapidly if it is placed in circulating cold water than if it is immediately placed in a refrigerator.

The production of high quality milk requires strict cleaning and sanitising procedures for all equipment that comes in contact with milk. Cleaning and sanitisation can be done manually or mechanically depending on the type of equipment used on the farm.

The basic steps in cleaning and sanitisation are as follows:

— Immediately after milking, or removal of milk from equipment, rinse the equipment with lukewarm water before the milk dries on the surface.

— Prepare a detergent solution according to manufacturers' specifications making sure the water temperature meets or exceeds the minimum recommended temperature. If manual cleaning is employed, brush all milk contact surfaces thoroughly. All milk contact surfaces that are not cleaned by mechanical cleaning or circulation cleaning must be brushed.

— Rinse detergent from tank with tap water. Preferably, an acidified rinse should be used to prevent the accumulation of milkstone.

— Drain rinse water from all equipment

— Immediately before using the equipment sanitise with an approved dairy sanitiser. Either chlorine or iodine can be used at concentrations of 200 and 25 ppm respectively.

54

Meat Goat Production

Goat is the most highly consumed meat in the world; and more goat's milk is consumed worldwide than cow's milk. The increasing economic importance of meat goat production can be attributed both to a strong demand for goat meat and to an interest in ecologically sound forms of vegetation control. Many ethnic groups – including Hispanic, Muslim, and Caribbean peoples – enjoy goat meat, called "chevon" by some and "cabrito" by others.

Demand is currently about double the domestic production, so there is ample room for expansion. Meat goats fit in well with other enterprises, particularly cattle operations, and may be used to control noxious weeds and brush to improve pastures for other livestock.

Meat goats can be raised with very little supplemental grain and with minimal shelter, and are generally an easy-care animal. The key management issues for a successful meat goat enterprise are fencing, parasite control, predator control, and marketing. Attention must also be paid to nutrition and to breeding stock selection.

While goats are enjoyable to raise and may be profitable, they are not a way to "get rich quick." As with any farming endeavour, knowledge and skills are essential for success. Prospective producers are well advised not only to read up on the subject, but to find and spend time with a local meat goat producer, and ask lots of questions.

Characteristics of Meat-goat Breeds

Goats have unique behaviours. They are intensely curious and will investigate anything that sparks their interest. Coupled with curiosity, their jumping and climbing ability can present some real management challenges. Goats can climb through a new "goat-tight" fence, pull the wash off the line, nip the rosebuds, or bounce onto a parked vehicle. Female goats are called does or sometimes doelings if they are less than a year old. Males are bucks or bucklings. Young goats are called kids. Bucks usually do not make good family pets because of their aggressive behaviour and strong urinelike odours.

Buck odours are most offensive during the breeding season, usually from September to early January. Does do not secrete strong odours from their scent glands. Goats adapt well to hot environments because of their small size and higher ratio of body surface area to body weight. Also, their ability to conserve body water, their limited subcutaneous fat cover, and their hairy coats are good survival traits under desertlike conditions. The foraging preferences of goats encompass a wider spectrum of plants than those of other small ruminants.

Goats are inclined to forage or browse from the top of a plant downward, making them an effective biological herbicide for controlling many undesirable plants and shrubs. Goats are called "selective browsers" because of their desire to choose from a large variety of vegetative types. This grazing behaviour enables them to survive harsher semiarid conditions than either sheep or cattle.

In some parts of the world, all breeds may be raised for fiber, meat, and milk and cheese production. Kids of all breeds can be used for meat. However, meat goat carcasses are generally leaner and more muscular than dairy goat carcasses, and have different proportions. Several meat-goat breeds are available in the U.S. The most widely available and the breed best suited to extensive range is the Spanish meat goat, also known as the "brush" goat. Most are horned; color and size are variable.

Figure 1. Spanish goat

Only horned bucks should be used, as naturally polled goats carry a gene for hermaphrodism. Spanish goats are characterised as hardy and adaptable, excellent foragers, and excellent mothers. However, their flighty disposition – if raised extensively – may make them hard to handle, and they are generally slower-growing and lighter-muscled than other types. Some lines of Spanish goats have been highly selected and will be far superior to the average.

Dairy breeds may be crossed with Spanish goats to produce a larger kid, and the resulting cross will produce more milk. However, the larger udders of the dairy breeds will cause problems in brushy areas.

Dairy goats' nutritional requirements during lactation are very high, and therefore more supplemental feed will be needed to maintain milk production. Dairy breeds are much calmer than Spanish goats. Because they have been selected for milk production rather than carcass qualities, dairy breeds will not normally produce a meaty carcass. However, their availability and price can offset the carcass characteristics, and crossbreeding with a Boer-type buck results in a desirable meat animal that is inexpensive to produce.

Angora goats may be raised successfully for meat. However, they are not adapted to cold climates, and are not as prolific as other goats. It is possible to raise them in northern areas – please refer to *Angora Goats the Northern Way*, by Sue

Drummond. The University of California Small Farms Center has a good article about raising angoras, Angora Goats A Small-Scale Agriculture Alternative.

Figure 2. Angora goats

Boer goats can be very expensive, but they grow more rapidly, put on more meat, and have a calmer disposition than other breeds. They are easily recognised by their large, muscular white bodies and red heads.

The Boer goat originated in South Africa, and was imported to the United States in 1993. Boer-Spanish crosses perform well, and using a Boer buck on a fine set of Spanish does is a good way to increase the muscle and growth of the kid crop without incurring excessive expense. Boers also does weigh as much as 200 pounds. They will therefore require considerably more feed than other breeds. Boer-cross does are said to be excellent mothers and good milkers.

In Montana, Boers have been crossed with cashmere goats, with excellent results. There is a high demand on the West Coast for these goats, and the fleece value (three to four dollars) offsets the shipping cost.

The Kiko is a New Zealand breed selected for survival and growth rate. They are large-framed goats, excellent mothers, and very hardy. The does can wean 45-pound kids with no extra input, and have a high twinning rate. Kikos may be expensive and hard to find. Excellent foragers bred under tough conditions, they are being used successfully in grazing-for-hire businesses. Contact Sylvia Tomlinson (Meat Goats of Caston Creek) or An Peischel, PhD, (Tennessee State University), to learn more about Kiko goats.

Tennessee Woodenleg goats, also known as "Fainting Goats" and "Tennessee Stifflegs," are myotonic – their muscles become extremely stiff when they are frightened.

Figure 3. Woodenleg Myotonic.

The attack usually lasts 10 to 20 seconds, and if they are off-balance when it hits they may fall over. This hereditary condition makes the Tennessee Woodenleg very muscular. The breed originated with four individuals brought to Marshall County, Tennessee, in the early 1880s, and the population of the breed is small.

Tennessee Meat Goats originated from the "Fainting Goat," but have been selected for heavier muscling and larger size.

Selection

Once finances, fences, abundant food sources (browse or pasture), and shelter are ready, it is time to acquire the goats. A small group of goats will provide many learning experiences in the first year or two. The group can easily be expanded as expertise is gained. Since one buck (male) can easily service 25 to 50 does (females), that is a logical herd size to begin with.

Of first importance is the health status of the animals, and it is a good idea to buy all your animals from one reputable breeder, if possible. Examine the entire herd, and be sure they have been managed the way you intend to manage them. Avoid limping animals and be sure to find out how the goats have been dewormed, and whether they have resistance to any dewormers. Other important features to check out before purchase are udders, teeth, hooves, and overall body structure.

A goat should not be fat. The hair coat should look healthy and shiny. Hooves, teeth, and udders should be sound. Teeth are important for grazing and browsing, and are an indicator of age. Avoid buying animals with broken teeth, or with wide gaps between the teeth. Be aware that each set of mature teeth indicates one year of age; therefore, a doe with four sets of large teeth (eight teeth) is already at least four years old. This should be considered when negotiating price. It's a good idea to examine the previous kid crop and to look at production records. Twinning percentage and kid survivability are important components of profitability.

Weaning weights are also important, and indicate milking ability of the herd as well as growth potential of the kids. Does may kid at one year of age, but producers may choose to grow them out instead of breeding the first year. A doe should certainly kid by two years of age, however.

Goats raised "extensively" may not have records. In that case, ask the producer about the kid crop and be alert for individuals with too much body condition (fat) relative to the others. The ones that look the best may be the ones that did not raise kids. Avoid those freeloaders!

Pasture Management

Goats, in general, should be figured at the rate of five to six does to one animal unit equivalent. If stocked according to area recommendations, they can be grazed along with other livestock. In marginal grazing lands, goats have been shown to complement both sheep and cattle.

Goats consume a higher percentage of brush and other less desirable plants; thus, they help to maximize the use of marginal pastureland as well as improve forage production over time.

A carefully planned rotational grazing programme can enhance pasture production and help control internal parasites. Deferred pastures and small grain pastures are good for kidding since they provide excellent feed for milk production. Supplemental grazing in stubble fields, corn fodder, winter rye or wheat pastures, and brassicas can be used either to extend the grazing season or to boost required nutrient levels for some critical phase of production.

Fencing and Facilities

Goats require tight fencing. A wooden picket (snow) fence can provide a temporary enclosure, but goats will eat through such a fence if it is used as a confinement structure.

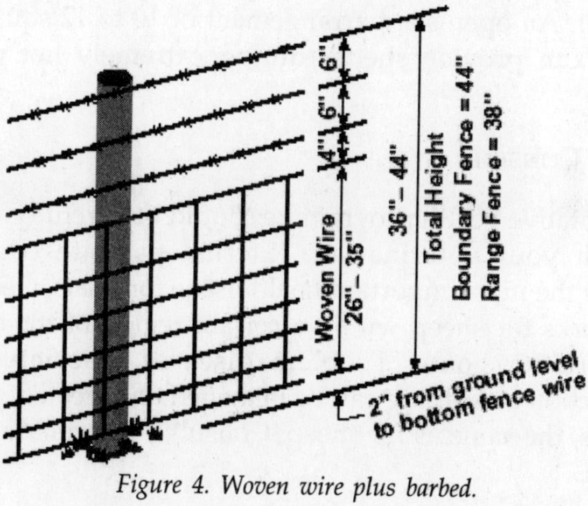

Figure 4. Woven wire plus barbed.

The more traditional woven wire fencing with two to three barbed wires above makes a good, tight fence (Figure 4). Likewise, the more recent eight-strand high-tensile fence with electrified third, fifth, and top wires has proven an effective goat-tight fence (Figure 5). Like other livestock, goats need some type of restraint facility and shelter when on pasture. Regular sheep-working pens are adequate for goats.

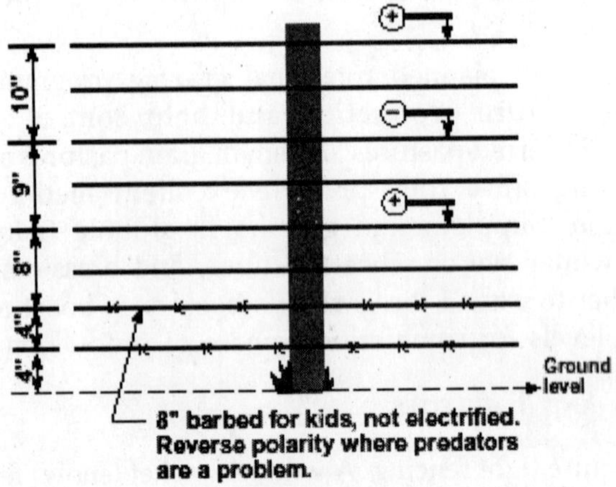

Figure 5. High-tensile electric fence wire.

Cattle pens can be easily adapted by making the lower section goat tight. An open shed arrangement of 10 to 12 square feet per doe can provide shelter during extremely hot or cold weather.

Health Considerations

A preventative health programme should be carefully worked out with your veterinarian. Internal parasite control is probably the most important health issue for goats. Generally, what works for sheep within a certain region of the country will work for goats. Problem diseases associated with reproduction or kidding can be managed and treated, in most instances, the same as for sheep. Consult a veterinarian when in doubt.

Marketing Meat

Kids are usually marketed at four to five months of age or before weaning. Options include direct marketing off the farm, supplying goat meat for speciality markets such as holiday or ethnic group uses, and producing kids for commercial marketing firms. Success is often a reflection of how well a producer tends to all aspects of breeding, health, management, and marketing.

All these factors have their respective roles in producing and marketing a quality product. Along with direct marketing to ethnic groups, there are two other potential niche markets for goat meat:

— target markets serving health-conscious consumers wanting low-fat diets, and

— the restaurant trade serving ethnic or gourmet foods featuring goat meat.

These markets are largely untapped and can provide real opportunities, especially for producers within the markets' immediate radius.

There is currently a strong and increasing demand for goat meat. Domestic slaughter and imports continue to rise annually, and goat meat that was once exported to Mexico, Canada, and the Caribbean is now being consumed in the U.S. The meat is lean, and may appeal to health-conscious consumers, but the primary purchasers of goat meat are members of ethnic groups, especially Hispanics, Muslims, and various Caribbean and Asian peoples.

The U.S. Census Bureau projects that between 1995 and 2050, Hispanics will account for 57 percent of the immigration into the U.S., and that Hispanics will account for 25 percent of the U.S. population by 2050. The vast majority of Muslims in the U.S. reside in the area stretching from Washington, D.C. to Boston, Massachusetts. Most of the Caribbean immigrants live in Miami, Florida, or New York City. There are strong Asian, Hispanic, and Muslim population on the West Coast. Peak demands for goat meat occur at Easter, on Muslim holidays, on the 4th of July, and at Christmas.

Farm Gate

If there are only a few kids to sell each year, and an ethnic population is nearby, "marketing" may mean simply finding one family who likes to have goat meat for the holidays or for a barbecue. Muslims may want to slaughter the animal on-farm for religious reasons.

Some Hispanics may prefer on-farm slaughter as well. If the producer agrees to allow this, it will be helpful to provide a few amenities. These may include a hose hooked up to running water, buckets, a flat working surface, and a hanging arrangement to suspend the carcass while skinning. Arrangements must be made for disposal of offal. Check state regulations for information about composting or burial.

In addition to sales at the farm gate, there are several channels for marketing meat goats. These include auction yards, private buyers, processors, sales to restaurants or grocery stores, and sales to a marketing cooperative. Prices at auction yards have improved in recent years, but are still dependent on the buyers present. A commission is charged, and the price is out of the seller's control. There is no advertising cost, and this may be the most convenient way to market the animals.

Private buyers may come to the farm or accept delivery at some other location. They will be re-selling the animals to consumers. The seller has more control over price, but perhaps less security in payment than with other methods of selling. Check out the reputation of the buyer, or ask to be paid in cash. Bargaining skills will be an asset in this type of transaction. In certain areas, restaurants and grocery stores with an ethnic clientele will be interested in buying goat meat.

Marketing Cooperatively

Individual producers may wish to organise into a marketing association to increase marketing options. Many buyers are more interested in a large uniform group of animals (lot), and will offer a better price for a lot than for an individual.

Whatever marketing options you pursue, offer a quality product and understand what the buyer prefers. For instance, if the buyer wants 45-pound kids, it will not pay to feed them to 80 or 90 pounds. Also, goats do not marble; extra fat is simply waste, and is very expensive to put on the goat and also expensive to remove from the carcass. To understand what the buyer wants – ask! Also refer to the ethnic calendar and specifications referenced above.

Breeding Stock

In addition to marketing meat, there is an opportunity to sell the highest-quality kids as breeding stock. This will require good records, research into what breeds are in demand in your area, and skill in advertising. It will also be necessary to have a realistic idea of what breeding stock are worth. Caution is highly recommended: the price of breeding stock should be related to the price of the meat animal.

Grazing for Hire

Besides meat and breeding stock, a third potential product of meat goat herds is the service of grazing for vegetation management. For example, Kathy Voth used goats to create fire breaks and help control forest fires in Utah and Colorado. Another goat grazing business is Ewe4ic Ecological Services, run by Lani Malmberg in Wyoming.

Goats can effectively control kudzu, leafy spurge, multiflora rose, knapweed, and many other problem plants. The goats reduce the need for herbicides; increase the diversity of pasture plants, especially grasses; add fertility to the soil; and are able to control weedy areas that are difficult to treat with other methods. For example, steep slopes on water reservoir dams, utility rights-of-way, and fire breaks near urban areas may benefit from the use of goats to control vegetation.

There are no standard rates; each contractor must investigate the individual situation and write a bid for each project, based on anticipated costs and desired profit. Major factors to consider before attempting to contract your goat herd for land reclamation or fire control services include:

— A competent and reliable herder to oversee the animals and make decisions on when to move them

— Sufficient temporary electric fencing to confine the animals to an area
— Dogs for herding and for guarding
— Transportation costs
— Costs of alternative (competing) methods – chemical, mechanical, or controlled-burn

A useful resource to learn more about the use of goats to reduce fire danger and for other applications is GOATS! For Firesafe Homes in Wildland Areas.

Multispecies Grazing

Goats make a valuable contribution to maintaining the productivity of the pastures they graze, particularly when used in a multispecies grazing scheme. Many cattle producers have found that adding goats to their farm or ranch increases profits while improving pastures.

Meat Goat Specialist Jean-Marie Luginbuhl of North Carolina State University believes that "you can add one or two goats per head of cattle without reducing beef production." Goats and cattle do not normally share parasites, and goats consume plants that cattle avoid. This increases the amount of grass available for the cattle, as the suppression of brush allows more grass to grow.

Stocking Rate

It is generally believed that six mature goats equal one cow on improved pastures and that ten goats equal one cow on browse or brushy areas. When grazing brush, it may be necessary to adjust stocking rates in order to accomplish your objectives. For example, when starting out with a very brushy area it might be desirable to stock two to four goats, or more, per acre. Later, as the brush disappears, some goats may need to be sold while a few are kept to control regrowth.

These figures depend on the carrying capacity of the land. Observation and adjustment are necessary. Some producers, including Mr. Jim Willingham of 8 Mile Ranch near Uvalde,

Texas, choose to allow the goats to harvest the brush as forage and maintain it as a renewable resource, rather than attempting to kill it.

Feeding Meat Goats

In order to raise goats at a low cost, the producer must maximize the use of forage. Feeding of goats cannot be discussed without mentioning the impact of the kidding cycle. Most goats are seasonal breeders, beginning to cycle with the shorter and cooler days of the fall. They will continue to cycle (unless they are bred) every 21 days or so, until days lengthen in late January or February. Since the gestation period is 150 days, this means that goats bred in September will kid in February, while delaying breeding until late November would mean that kids arrive in late April and in May.

The time of kidding determines the period of highest nutritional demand, as late pregnancy and early lactation are critical times for the doe and kid. By manipulating the breeding date, the producer can see to it that those peak needs hit when more forages are available, rather than during months when only harvested feed can be used.

Table 1 : Dietary Protein and Energy Requirements of Goats

Class of Goat	Ave. feed intake/day lb	% Crude Protein	% TDN
Growing doeling, 45 lb	2.4	8.8	56
Growing male kid, 66 lb	2.9	9.0	57
Yearling doe, 90 lb	4.6	10.0	56
3 yr. old doe, 110 lb	5.0	11.7	69
Mature buck, 220 lb	5.3	9.0	55
Dairy doe, 150 lb	7.5	11.6	71

"During the warm-season grazing period," write the Pinkertons, "goats will very likely meet all their nutritional requirements from whatever combination of forages is available; only a trace mineralised salt and possibly some phosphorus would be needed in addition." However, in late fall and winter you will probably need to supplement.

Table 2: Practical Dietary Recommendations for Feeding Goats

	% protein	% TDN
Growing kids, dry does, and bucks	9-10	54-58
Growing kids, dry does, and bucks	10-11	56-60
Lactating goats	12-13	62-68

Here are some options recommended by the Pinkertons, based on their experiences:

— Provide a few hours of grazing on ryegrass or small-grain pastures.

— Offer grass hay ad lib plus 1 pound of 20 percent protein pellets daily.

— Feed higher-protein hay (12-13 percent) ad lib, and provide one-half pound of corn per head per day for pregnant or lactating does.

— For kids 3-6 months old, use 1 pound of 16 percent protein feed plus grass hay. Older kids can have grass hay plus 1 pound of 14 percent protein feed.

Using these figures, a producer can estimate feed costs. For example, if you plan to begin with 25 does, and in your environment it is typical to feed hay for the months of December, January, February, and March, then it is possible to calculate feed needed for the season:

120 days X 26 goats (does plus buck) X 5 pounds (approximate dry matter required) = 15,600 pounds.

This is on a dry matter basis. Convert that figure to as fed basis by dividing by 0.9, since hays are typically about 90 percent dry matter.

15,600 divided by 90% = 17,333 pounds of hay.

Goats are notoriously wasteful; add on about 20 percent to allow for waste, which brings the total amount of hay needed to approximately 20,800 pounds, or 10 tons. If bales are 60 pounds each, then each goat will be eating about 13 bales over the course of the winter.

Next, calculate the number of pounds of protein supplement that will be needed for the year.

120 days X 26 goats X 1 pound/day = 3,120 pounds protein supplement for the winter season.

Convert that figure to a cost as well. If energy supplementation is needed as a result of the type of hay used, calculate the number of pounds needed, and the cost.

There will be other costs, such as fencing, pasture expenses, salt, and minerals. Individual situations will vary tremendously when it comes to fencing and pasture expenses, depending on whether or not there is an existing fence that can be modified inexpensively. If ample browse is available, there may not be any pasture expense.

When feeding goats it is very important to observe closely and adjust feeding practices based on how the animals are doing. A ration that looks adequate on paper may turn out to be unpalatable, or may need to be increased due to severe weather conditions, or may be overly generous if the goats are finding plenty of browse.

A properly nourished animal will be healthier, and more able to handle stress and bad weather. An over-fat animal will have a whole set of problems, and will be a drain on the budget as well. Some areas of the country need selenium supplementation.

Profitability

One of the key questions to answer before starting an enterprise is, "Will it be profitable?" The answer is largely dependent on the management and the set of individual circumstances. Many sample budgets have been published, and they are useful to help sort out the various categories of expenses that must be considered. There are some basic principles to keep in mind that will improve the chances for profit. There are four fundamental conditions for success in a livestock enterprise:

— Must have a viable market for your product.

— Market price must exceed cost of production.

— The goal for reproductive performance is at least one merchandisable unit per exposed female.

— Match genetic potential for growth with productivity of the environment.

Machen includes a table to illustrate the relationship between cost of production and reproductive performance. Clearly, the chances for profitability are far better if costs are kept low and does are productive and kids survive.

— Further suggestions for improving the odds of success include:

— Start with good-quality, healthy breeding stock.

— Avoid high-dollar stock until you have gained experience. This lowers risk as you will inevitably make many mistakes during the first year or two.

— Keep expenses to a minimum. This is accomplished by using forages as much as possible, keeping equipment simple, and using the services of a veterinarian to set up a preventive health care plan.

— Maximize income by maximizing the number of animals for sale. This means concentrating on reproductive efficiency (kid crop born) and on keeping the kids alive and well through good nutrition, health care, and predator control.

55

Commercial Goat Farming

Goats play a significant role in ensuring livelihood security to the millions of small and marginal farmers, landless labourers and rural folk. Goat rearing under intensive and semi-intensive system for commercial production is gaining momentum. A number of commercial goat farms have been established in different regions of the world.

Goat rearing is an enterprise which has been practiced by a large section of population in rural areas. Goats can efficiently survive on available shrubs and trees in adverse harsh environment in low fertility lands where no other crop can be grown. In pastoral and agricultural subsistence societies in India, goats are kept as a source of additional income and as an insurance against disaster. Goats are also used in ceremonial feastings and for the payment of social dues. In addition to this, goat has religious and ritualistic importance in many societies. The advantages of goat rearing are:

i) The initial investment needed for Goat farming is low.

ii) Due to small body size and docile nature, housing requirements and managemental problems with goats are less.

iii) Goats are friendly animals and enjoy being with the people.

iv) Goats are prolific breeders and achieve sexual maturity at the age of 10-12 months gestation period in goats is short and at the age of 16-17 months it starts giving milk. Twinning is very common and triplets and quadruplets are rare.

v) In drought prone areas risk of goat farming is very much less as compared to other livestock species.

vi) Unlike large animals in commercial farm conditions both male and female goats have equal value.

vii) Goats are ideal for mixed species grazing. The animal can thrive well on wide variety of thorny bushes, weeds, crop residues, agricultural by-products unsuitable for human consumption.

viii) Under proper management, goats can improve and maintain grazing land and reduce bush encroachment without causing harm to the environment.

ix) No religious taboo against goat slaughter and meat consumption prevalent in the country.

x) Slaughter and dressing operation and meat disposal can be carried without much environmental problems.

xi) The goat meat is more lean and relatively good for people who prefer low energy diet especially in summer and sometimes goat meat is preferred over mutton because of its "chewability"

xii) Goat milk is easy to digest than cow milk because of small fat globules and is naturally homogenised. Goat milk is said to play a role in improving appetite and digestive efficiency. Goat milk is non allergic as compared to cow milk and it has anti-fungal and anti bacterial properties and can be used for treating urogenital diseases of fungal origin.

xiii) Goats are 2.5 times more economical than sheep on free range grazing under semi arid conditions.

xiv) Goat creates employment to the rural poor besides effectively utilising unpaid family labour. There is ample scope for establishing cottage industries based on goat meat and milk products and value addition to skin and fibre.

xv) Goat is termed as walking refrigerator for the storage of milk and can be milked number of times in a day.

Goat make a valuable contribution to the livelihood of economically weaker sections of the society. Amongst the livestock owners goat rearers are the poorest of the lot. Realising the importance of goat in the agrarian economy of the country, various developmental activities have been taken up by Govt.of India. The Central Government had established Central Institute for Research on Goats at Makhdoom, Farah, Mathura District,UttarPradesh. During VIII Plan Period Seven Intensive goat breeding farms were proposed with the objectives:

i) To produce 1000 stud bucks per year for the distribution among goat rearers

ii) To improve yield of milk and chevon through selective breeding of regular breeds like Jamnapari, Beetal, Barbari, Jakhrana, Jhalawadi.

iii) Cross breeding of non Pashmina goats with Angora goats to produce Mohair in Jammu and Kashmir, Himachal Pradesh and Uttar Pradesh.

iv) To conserve germ plasm of regular breeds like Jamnapari, Beetal, Barbari, Black Bengal, Malbari, Sirohi etc.

Goat Breeding Practices

Female kids will come into season in their first autumn (September to December), and could be mated at this time i.e. at about 6 months old provided they are well grown and mature but the aftercare of these "kidded kids" needs to be especially good if the goat's full potential is to be realised. In the UK, most pedigree breeders do not mate their female goats until they are about 18 months of age. Male kids may become sexually active when only a few weeks old! This means that male kids need to be kept separate from females from about six weeks of age or there could be "accidents". Male kids are used for stud purposes in their first autumn, although late born kids may not be ready until later in the breeding season.

The sure signs of oestrus are wagging of the tail with a rapid side-to-side movement and wetness around the vulva and the feathers of the tail. Other signs are redness and swelling around the vulva, noisy bleating and a change in behaviour. Not all of these signs may be present, and the wagging may only become present when the female smells the male. If you run your hand along the female's back towards the tail end, wagging may be stimulated too. If you have a male on the premises, you can use him as a teaser i.e. close, but not too close. Alternatively a "Billy rag", which is a piece of cloth smelling of male goat (normally kept in a sealed container!), can be waved in front of the female. Once a goat starts coming in to season, if unmated it will generally return every 21 days. The length of season is generally two or three days but can last from a few hours to four days or more. Shorter seasons are more common at the beginning and end of the breeding season.

Two methods of breeding are commonly used in goats: natural breeding and artificial insemination. Each form of breeding has its own advantages and disadvantages.

Natural Breeding

In natural breeding, a doe (female goat) in heat is bred to a buck (a male goat).

Advantages: Goats have been breeding naturally for centuries. As an owner of a doe, you must decide which buck to use, but the buck takes care of the rest.

If you don't own a buck, many goat owners will allow you to bred your doe to their buck for a fee. In our area, this fee ranges from around $35 to $100 depending on the quality of the buck. Most also charge a fee of around $1 a day to care for your doe if you must leave it to wait for it to come into heat. If you own a buck, your costs include the purchase price and the cost of feeding and housing the buck. You can make some money breeding other's does if you do own a buck.

A doe is in heat from 12 to 36 hours, and after breeding naturally semen can live for about 24 hours. It's not essential to bred at exactly the correct moment, and it's easy to repeat the breeding for a few days.

Disadvantages: Difficulty in managing a buck: Bucks require sturdier fences than does, and are usually housed away from does. During breeding seasons, bucks have some unusual behaviours such as urinating on their beards. This creates an offensive smell that attracts does but only does seem to appreciate it.

If you own your own buck(s), you'll have only a few to choose from. If you use a buck that belongs to someone else, you'll still have a limited supply to choose from within driving distance. If you don't have a trailer or pickup truck, it can be inconvenient driving your doe to be bred.

Bucks frequently cost more than does, and you'll probably have to own several because you usually won't want to breed the offspring of a buck to the buck. Bucks also eat more than does, and destroy their fences more frequently.

Artificial Insemination

In artificial insemination (AI), frozen semen from a buck is defrosted and inserted into the cervix of a doe in heat. Goats conceived by AI look and act just like other goats.

Advantages: Once you have the necessary equipment, frozen semen is much less expensive than paying a breeding fee. Typical prices range from $5 to $25 per "straw" of semen. You can also avoid the costs of housing and feeding a buck.

Straws are available from many high quality bucks, including National Champion bucks, and sires and siblings of National Champion does. It's possible to quickly improve the quality of your herd using such resources.

Disadvantages: The best time to AI a doe is right at the end of the heat cycle. Defrosted semen remains viable at most a few hours and is usually immediately after defrosting. Therefore, it's important to know exactly when the doe went into heat and how long she remains in heat to be successful.

The necessary equipment includes a liquid nitrogen tank, "straws" of semen, tweezers (for removing straws from the tank), a straw cutter, (for cutting off the wax plug keeping the semen in the straw), an open ended glass speculum, a small light (to make the cervix visible), a thermos (for holding warm water to defrost the semen), a thermometer (for measuring the temperature of the water), an insemination gun (A thin metal tube with a plunger to push the semen out of the straw) and disposable plastic sheaths (to hold the straw while in the gun).

Choosing Correct Breed Type

Male goats should be of correct breed type with sound conformation. Male goat shows are often a good place to observe potential sires, whilst a visit to see the male in his home environment, alongside relatives and daughters is highly recommended. In dairy goats, more so than with any other type of livestock, studying the pedigree of your goat and its potential mate gives valuable information about the performance of their ancestors, as milking awards and milk yields form part of each animal's permanent record. Breeders often use a pedigree a minimum of five generations, because very close breeding, in the fifth and later generations, has a very strong influence on the pedigree of potential offspring.

Livestock Management

From birth, pregnant till kidding, the management is divided likewise

- — Birth till weaning (3-4 months)
- — Weaning till pregnant
- — Adult goat management
- — Management of pregnant doe until kidding
- — Management of breeder buck
- — Management during flattening stage until sale
- — Birth to 3 months.

Kid Management

Within 6 hours of birth, kids are given colostrum . Colostrum are excreted for 3 days by the doe. Usually kids suckle without assistance, if not, the doe is milked and the kids are bottle-fed. Colostrum intake is very important for antibody production to fight against diseases and it also initiates immune response.

Kids' navels are dipped in iodine to hasten drying and it also prevents infection. Kids depend on mothers milk from 3 – 4 months. Beyond 2 months milk intake decreases and the kids start grazing a little.

After 3 - 4 months, kids are weaned and start grazing. Weaning is done in stages, whereby the concentrates, grass and milk are given during this conversion stage. The doe and kids are kept in the pens. After 4 - 6 weeks, they are grazed in nearby areas for easy monitoring.

Pens for kidding doe and kids should be clean and dry. Damp pens are chilly. As kids are sensitive to cold and thus may easily contract pneumonia. If colostrum are not given, a mixture of milk with a bit of vegetable oil and egg white can be used to stimulate intestinal peristalsis.

Kids can be fed from other does' milk after milking or suckled directly from other does if they permit. Powdered milk can be mixed and given. Cows' milk can also be given by mixing with 1-2 drops of fish oil at each feed. About 170 grams of milk are given (less than a cupful) every 4 hour (QID). This amount can be increased as the animal gets older.

Post-weaning Till Adult

After weaning at about 3-4 months, kids are separated from the does. They remain separated in the pens and grazing pastures. Feed intake are monitored closely as drops in bodyweight commonly occur after weaning due to stress from separation from the doe Male and female kids are separated as goats start breeding at 9 months. Kids not meant for breeding can be sold

Adult Goat Management

Management of adult goats are the same as for kids from 4 months to 1 year of age.

Pregnancy and Post-kidding

Pregnant females and those just kidded must be separated from the kids to avoid accidental trampling of the newborn. This simplifies management. Pregnant does require close attention a month or a forthnight before and after kidding. Good quality feed is provided. This is to ensure healthy mothers and kids. Clean jute sacks can be placed on the floor and the pens' sides to protect kids from the draft.

Breeder Bucks

Male and female are kept in separate pens. They are only brought together for mating. All males are kept in separate pens to avoid fighting and injury. Each male may serve 20 females.

All potential breeder males must be well looked after in terms of their nutrition and health. They need a lot of exercise and movement. If insufficient, the animal can become lazy and lethargic. They should have good strong legs and have their hooves trimmed regularly, as overgrown nails can hinder mating activities and cause health problems. Male kids can mate and breed at 9 months, but they are best used at 1 year of age.

Fattening Management

It is easy to farm goats for fattening and sale for chervone. The only important factor is feed which must be high in nutrients and of good quality.

Kids from 6 months of age can be placed in the fattening scheme. Fattening needs 6 months to 1 year to achieve a bodyweight of 20-25 kg. This is base on the type of feed and ability of goat to gain weight. One to 2 months before sale, they should be given quality high protein feed.

Management of Sick/Injured Goats

All sick or injured goats must be separated from the group and given immediate treatment.

General Management

Area of management	*Explanation*
Area	The cleanliness of the pen area and surrounding is important. The must be good drainage around the pen and area. Droppings are discarded away from the pen. Pature for grazing or cut and carry must be fenced, to make grazing easier and to prevent mixing to prevent accidental mating that might arise
Labour	All work must be systematic and simple. Whilst wating for the goat to be let out for grazing, clean all feed troughs and cut the grass. After goats are let out to graze the stalls are cleaned
Time	Daily husbandry and long term management must be planned from time to time. Planned systematic management are like breeding, kidding, dehorning, castration, deticking, deworming. Other management are like paddock roation, fertilizing, as the used of machinery, labour and other implements can be done routinely.
Identification	Every kid born must be given a record card. Identification by ear tagging or tattooing must be done
Separation	Separation of males not intended for breeding is done at the age of 4 – 5 weeks old to promote faster growth. This is beneficial for fattening and can increase effiency of breeding management

Dehorning	This is done during the kid stage to minimise injury caused by fighting or accidents. Horn that regrows must be removed.
Hoof trimming	Goat that are penned for long durations develop long hoofs that cause difficulty in walking and can be a source of infection
Hair Trimming	Crossbreds have long hair that need to be cleaned. Soiled hair can be a source of infection and breeding place for lice. Hair around the udder are trimmed for easy milking and prevent any hair from getting into the milk
Deticking	Deticked by spraying or dipping

Goat Farming Systems

Intensive Systems

Intensive systems needs good housing as goats are not let out to graze. Feed and water provided daily- concentrate, grass and salt lick. This is practiced by farmers with limited land space or for pregnant doe and those that had kidded and also for breeder buck

In Intensive systems chances for infection are high. Stalls must always be clean and dry. This also make monitoring of goats for treatment and disease suveillance easy.

Free Grazing System

In free grazing systems, goats have large grazing areas. Goats graze full time. Management is difficult – animals tend to wander. Grazing systems combined with good grazing management skills have long enhanced overall production of livestock operations. Correct stocking rate is critical to the success of any grazing system. Before applying a grazing system to your operation, you must determine what system best fits your operation's goal and what level of management is available to operate it.

Semi Intensive System

In semi-intensive system animals are let out to graze in the day and are stalled at night. The benefits are they are protected at night and during bad weather, it also gives them liberty to graze better quality grass and herbage This also gives opportunity for kids to exercise and get enough sun. Health checks can be done when the animal are brought in the evening.

Housing of Goats

Choosing an Area

Choosing an area for farming is important. Area chosen should be:

- There should be big trees for shade and to act as wind breakers
- There should be a mixture of hilly and low areas with good irrigation and good air movement
- Far from busy highways

The siting of the housing should take into consideration:

- Minimise distances for movement during feeding. Example, grazing area/feed store must be near the stalls
- If it is required to build more than one house, the distance between the houses must be at least equivalent to the width of the stall. This is to prevent infections .
- Good ventilation is important to cool the air flow in the shed
- To allow as much morning sun into the shed. The shed is in the North-South Orientation
- Do not get damp easily. Good drainage system around the goat houses
- Do not get strong winds
- Protected from rain and heat
- Waste/droppings easily collected and disposed

— Sufficient room for feed and water
— Suitable flooring which does not cause injuries such as sprains and fractures which may result in the spread of diseases.

Building the Shed

The shed can be on the ground or a raised floor system. Raised flooring is recommended as it provides good ventilation and is easy to keep clean

Flooring

The flooring should:

— hardy, for example should withstand chemicals and urine
— easily cleaned
— not slippery
— does not cause injuries
— does not get damp easily

The flooring should be made of strong material such as wood measuring 1.5 cm (thickness) and 3.5 cm (width). Split bamboo or split trunk of areca nut tree, or branches and normal wood. For excrement and urine to fall through below, the planks or wood are spaced 1.2 – 1.8 cm apart; but not too far apart to prevent legs from getting stuck

Walls

Walls must be of hardy material. This is due to the goats' habit of ramming their horn, biting or rubbing against the wall. Recommended material are bamboo, round wood and planks. The wall is divided into two parts: from the floor to about a meter up it may be made of planks, with a gap of 1.5 – 2.0 cm. Above this, strong wood, bamboo or sago, branches can be used.

Roof

Materials that may be used for roofing include:

- *Sago palm fronds (Nipah)* - good for ventilation and cooling. (generally, goats have been seen to eat the sago fronds)
- *Zinc* - easy to assemble but can cause the shed temperature to soar, especially if there are no shade trees around. The height of the roof must be increased to provide good ventilation
- *Asbestos* - the best material for roofing but costly. Its not only cooling but also do not burn easily
- *Aluminium* - reduces effects of heating but costly

Feed Troughs

Feed troughs can be built outside or inside the stalls, along the centre of the shed. It is 0.3 m wide at the bottom and 0.45 m wide at the top. For stalls without feeding trough, grass are tied about 0.5 cm above the floor. Enough water should be provided in buckets or basins.

Feeding of Goats

These are needed for growth, breeding and health. It affects the profitability of the livestock enterprise. Green feed has high moisture content and it is a source of energy, vitamins and minerals. Concentrates has high dry matter content, is rich in protein, fat, mineral and energy. Insufficient and unbalanced feed may cause diseases in livestock

A balance meal contains the following nutrients:

Nutrient	*Information*
Proteins	Building strong muscles and for tissue repair
Carbohydrates	growth
Vitamins	Required only in minute amounts but it is essential. If not balanced it can lead to diseases and the animal is more susceptible to infections

Minerals	Minerals are essential for bone, teeth and health. These minerals include calcium, phosphorous, sodium. E.g.: ordinary salt, salt lick, calcium phosphate, bone meal, fish meal
Water	This is also essential in feed, not only as a thermoregulator, but also for mineral transport and helps mineral intake in the body.

Grass and Legumes

Goats have a reputation for being able to survive and produce on low quality, brushy, and/or woody feed. In fact, goats require more nutrient rich diet than most other ruminant animals. However, because of the way its mouth is constructed together with its unique foraging preferences, a goat will often consume the nutritious parts of species that are otherwise woody and low in quality. Nevertheless, both cool- and warm-season grasses and legumes can provide much greater yields of high quality forage for goats than weeds and brush.

One of the greatest concerns of the goat grazier is often uncertainty over the consistency of production and quality of pasture forage. A key strategy to address that concern is to include a diversity of forage species in the grazing system. Diversity can exist both within and among pastures on the farm. Every species has a somewhat different distribution of growth through the year and site/climatic adaptation; including a diversity of species will ensure that something is green and growing during most months of the year.

There is no "silver bullet" species for a grazing system. No single species will be adapted to the entire farm, nor will it provide satisfactory production throughout the year. The mid-Atlantic region is fortunate in that a large number of both cool- and warm-season species are well adapted.

Legumes are one of the most critical components of a sustainable grazing system. The grazier should have a goal of maintaining 30-40% legume in most of his cool-season pastures. The only cases where legumes may not be desirable are with warm-season grasses or with tall fescue that will be stockpiled to provide winter grazing. (Stockpiled tall fescue pastures that have a strong legume component should be grazed by late fall because legumes deteriorate rapidly after hard frosts.)

Legumes are critical for 3 major reasons. First, they fix atmospheric nitrogen which supplies nitrogen for their own growth in addition to grasses growing in association with them. This is important because it reduces dependence upon commercial N fertilizer with its associated costs and environmental concerns. Legumes are also generally of higher quality than grasses. While protein is the most frequently cited advantage of legumes when compared to grasses, more importantly, legumes have greater intake potential than do grasses. Third, legumes, and in particular alfalfa, help to stabilise the uniformity of pasture production during summer months when cool-season grasses are often unproductive.

The goat grazier should also endeavour to maintain the majority of his/her pasture in a strong perennial grass sod. Sod-forming grasses include species such as bluegrass, fescues, reed canarygrass, and smooth bromegrass. These grasses form rhizomes (underground stems) which enable them to form a turf that is beneficial for a number of reasons. A sod-based pasture protects the soil from erosion as well as compaction. It is especially critical that poorly-drained soils have a good grass sod to protect against soil compaction by animal traffic. The sod also helps to anchor crown-forming legumes in the soil.

While often scorned, tall fescue is a desirable forage base for the goat grazier. Tall fescue is a tough, low-maintenance grass that can provide good quality forage if kept in a vegetative stage. Its excellent performance under stockpiling for fall and winter grazing together with its winter tolerance to hoof traffic are two of tall fescue's outstanding features. While much of the existing acreage of tall fescue is infected with a fungal endophyte that reduces animal performance, maintenance of tall fescue in a vegetative stage and/or dilution of the stand with legumes are two practical means for minimising the endophyte's effect.

Orchardgrass is probably among the best choices for a cool-season grass in a perennial pasture. Kentucky bluegrass can be seeded along with orchardgrass to help to provide a good sod base. Many graziers find that bluegrass increases naturally as they move hayfields into rotationally-stocked controlled grazing situations. At high elevations or in the more northern portions of the mid-Atlantic, Kentucky bluegrass may provide acceptable production as the only grass species in a mixture with white clover. Smooth bromegrass is a taller-growing grass that also is a good pasture grass in more northern areas. Reed canarygrass is among the cool-season perennial grasses adapted to the broadest range of soil conditions. While it is frequently recognised for its adaptation to wet, poorly drained areas, reed canarygrass also has greater tolerance to hot, dry summer conditions than any other cool-season grass.

Grasslands Matua prairiegrass or rescuegrass is a Brome species that is most commonly referred to simply as Matua. This variety was developed in New Zealand in the 1970's and has been heavily promoted in the US in the 90's. Matua had shown good potential in many grazing situations. It is very responsive to applied N, and has also done well in mixtures with alfalfa. Key to maintaining perennial stands of Matua is allowing seedhead maturation and seed drop at least once per year. Some Matua plants can overwinter, but many do not, so natural reseeding is required to maintain the stand.

Alfalfa, although typically used for hay and haylage production, is an excellent legume for pasture with controlled grazing. It is best mixed with a tall-growing cool-season grass. A number of alfalfa varieties specifically marketed as grazing-type alfalfas are now available. Under well-managed grazing, where defoliation occurs within a few days and plants are allowed adequate time to regrow and replenish reserves before being grazing again, hay-type alfalfa varieties may be just as effective. The most important considerations with alfalfa are to 1) choose a variety adapted to your area with good disease resistance, 2) establish only in areas with a pH above 6.2, at least medium levels of P and K, and not poorly drained, and 3) use rotational stocking. Not observing any one of the above recommendations will seriously limit alfalfa's productivity and persistent and will very likely result in a lot of money wasted.

White and red clover are both excellent legumes for grazing. These legumes should always be grown in association with grasses when being grazed. Periodic broadcast seeding of red clover and sometimes white clover will be necessary in most situations. Red clover is a short-lived perennial that rarely lives more than 2 or at most 3 years. White clover can live for much longer periods, but is very sensitive to stand loss during hot, dry weather. Clovers offer some advantages to alfalfa in grazing situations including 1) not susceptible to alfalfa weevil, 2) less sensitivity to soil fertility and drainage status, and 3) favourable establishment with frost seeding.

Cereal rye and ryegrass offer opportunities for grazeable forage during late fall, late winter, and early spring. These species can be no-till seeded separately or together into warm-season grass stubble. Both perennial and annual warm-season grasses can be used for grazing. Switchgrass and caucasian bluestem, eastern gamagrass, and bermudagrass are perennial, warm-season grasses that provide remarkably reliable forage of good quality for grazing during the hot, dry summer months. Annual warm-season grasses include sorghum, sudangrass, foxtail or dwarf pearl millet, grazing maize, and even crabgrass. These species provide reliable grazing once established and can be no-tilled into ryegrass or cereal rye without herbicide suppression or into a cool-season sod after herbicide burn-down in late spring-early summer.

Supplementary Feed

Supplementary feed is important during gestation, milking and does with a newborn kid. Pregnant Does or high milk producers is given more concentrates (½ - ¾ kg. per day), whereres dry mothers and growing kids need about ¼ kg. per day. At weaning when stress due to weaning and adjustment to new feed, kids can be given supplementary feed in stages from 1 – 15 months

Feed for strengthening are from fine concentrate, maize, coconut pellets, groundnut pellets, soya waste and others. As a guide, these extra feed can be a mixture of concentrate and agriculture waste.

Improving Milk and Meat Production

Many factors can have major or minor effects on the goal of improving milk and meat production of goats. Maximisation of production is not guaranteeing maximisation of net return. The reason is that the law of diminishing returns governs much of animal production, especially feeding and the relationship to reproduction. It means that for additional units of input, such as feed, there is an ever-decreasing increment of benefits in units of milk and meat dollars or other output, until certain general bases and levels are lifted for a new set of overall conditions. It also means that additional feed will produce additional pounds of milk and meat up to a certain biological limit, but in the meantime may already produce negative income dollars from a certain point on.

Other important factors influencing production output and income from goats are genetic merit, udder quality, health and marketing, besides feeding. Genetic merit comes from:

— selection of native goats,

— crossbreeding with improver breeds.

Selection of native goats can be very effective, because of the inherent adaptation to the climate, especially if it is tropical, hot and humid, and the resistance to native diseases, insects and parasites. Selection requires regular record keeping of each herd animal in terms of production traits, milk, composition, meat, growth.

Crossbreeding has the advantage of selecting presumably superior genetic producing ability, but adaptation to climate, diseases, insects and parasites is usually a big, often insurmountable or at least very expensive problem, which may only be solved by using for continued breeding crossbred offspring rather than purebred parents. In either case it is necessary to realise that improved feeding is wasted if there is no simultaneous genetic improvement of the basic producing ability, because

— heritability of milk yield by goats is about 25 percent,
— heritability of goat milk composition about 50%, and
— heritability of goat weight gain about 40%.

Heritability values can be used to predict the expected average progress from selection, assuming that environment, management, feeds and feeding, and climate is not changed between generations. Improvement in milk yield is often the most profitable choice and the predicted progress would be per generation:

heritability X selection differential or

25% X (milk yield of selected sire - milk yield of dam).

For example, if the selection differential from the buck proof is + 400 lb and the milk yield of the doe is 1,500 lb, then the expected average genetic improvement in the performance of the offspring in the next generation would be

(25% X 400 lb) + 1,500 lb = 1,600 lb

indicating that genetic selection is important, but 75% of milk yield performance progress is due to management, environment, diseases, climate and especially feeds and feeding.

The udder is the most important part among the inherited physical capabilities of the animal in its body parts and constitution. For centuries, this was one of the principal goals of attention of Swiss goat breeders, to improve the udder quality and conformation, and they accomplished this without parallel, making the Swiss dairy goat breeds the milk production leaders in the world. Today type evaluation is available, called the Linear Appraisal System, which can effectively aid in the selection for and improvement of goat milk production.

Health is the other important factor for success in goat management. Elevated, slotted floor barns have become popular in humid and hot climates for better health of goats, especially for internal parasite control. Such barns are easily and cheaply constructed, provide cool shade and dry areas for feeding and rest, they keep udders clean and free from contaminations and infections, and they prevent reinfestation from internal parasite eggs in feces, because the goats are resting on the slotted floors away from their feces. Without such basic provisions for optimum health of goats any attempts in feeding improvement are wasted.

Feeding for Health

The feeding program needs to aim for more than just higher milk yield or weight gain; it needs to provide the best possible health also through feeding, because this will directly affect readiness and success in reproduction. It has been said often that the goat has been neglected in research and numbers of publications, but this was true only until 30 years ago. Meanwhile there has been a ground swell of efforts recognising the goat as an important part of agriculture, especially small holder agriculture, and in the production of valuable food for human needs for self-sufficiency, diversification, risk stabilisation, natural resource utilisation like no other animal, gourmet foods and for people with medical needs like cow milk allergy, digestive malabsorption and cholesterol problems.

Fiber

A major concern in feeding for better health is the problem of enterotoxemia or overeating disease. Many goat managers vaccinate against it successfully, to prevent the associated toxin produced by Clostridium perfringens. Actually, enterotoxemia is caused primarily by acidosis in the rumen due to faulty feeding. At any age, symptoms of diarrhea, depression, incoordination, digestive upsets, coma and death are observed after excessive feeding of kids or mature goats, when sudden changes of feeds occur, when goats are hungry and had free access to palatable, readily fermentable feeds, when goats are fed too little calcium supplement and when too little roughage with too short fiber is fed. The best prevention for nursing kids is to have frequent feeding or nursing immediately starting after birth, so that kids are never hungry. Large meals, once a day and of little variety should be avoided. Goats are by nature browsers and like to select various feeds. High levels of grain feeding relative to roughage in the ration (> 60%), especially in early lactation lead to rumen acidosis, followed by inappetence and indigestion. Feeding buffers like sodium bicarbonate and magnesium oxide and stemmy hay will help alleviate the early symptoms and prevent enterotoxemia. Other effective feeds are sunflower seeds, cottonseed, oats, dry brewers grain.

Fiber is a feeding requirement unique to ruminants, because:

— it maintains a beneficial rumen flora, that produces mainly acetate, the important energy source for all ruminants, rather than propionate from starch fermentation; and

— it causes extensive regurgitation for rumination and plentiful salivation for rumen buffering, rather than fast passage through the rumen and incomplete digestion.

Fiber is rarely stated in nutrient requirement tables, but from dairy cattle research it is recognised, that at least 17 percent of the daily dry matter intake is needed. However, it makes a difference whether this fiber is shorter than 1 inch or longer. Effective fiber needs length to stimulate chewing and rumination. Feeds and diets, which cause significantly less chewing are potential problems leading to acidosis and enterotoxemia. When less frequent chewing is observed and before other more serious symptoms occur, a drop in milk fat content of 1 to 2 percentage units will be noted in a few days, the socalled low fat syndrome.

Vitamins

Feeding ruminants and their rumen microflora correctly should result in sufficient amounts of rumen synthesised B vitamins. However, any change and upset in feed intake may reduce the amounts significantly.

Niacin is a water-soluble B vitamin functioning as coenzyme in energy metabolism and is needed by high performing dairy animals, especially in early lactation when ketosis or acetonemia may be a problem. Some supplementation has been beneficial, especially when there is much corn in the ration, since corn is low in the amino acid precursor for niacin, and niacin deficiency may develop.

Thiamine deficiency may occur after heavy grain feeding or if certain feeds with antithiamine activity are ingested. Blindness can result and thiamine supplementation may be needed.

Pyridoxine is another B vitamin synthesised in the rumen and is required for biosynthesis of fatty acids, transport of amino acids and minerals. Upset rumen metabolism can lead to deficiency in pyridoxine synthesis and symptoms of anemia in the dairy goat. Not all anemia in goats is necessarily due only to internal parasites, but not much research into vitamin requirements and metabolism of goats has been done in recent years.

There are other feeding related disorders in goats, which are preventable. When goats are becoming fat at the end of lactation, they risk getting acetonemia, pregnancy toxemia or ketosis problems at or soon after kidding. Bearing triplets or quadruplets can aggravate the condition triggered by hypoglycemia. Prevention of undue weight gain in the dry period is often easier than treatment and correction when symptoms of dullness, depression, acetone odour in the breath, recumbency occur, which can lead to death. Gradually increasing feeding of 1 to 1.5 lb of concentrates 3 to 4 weeks prior to kidding is usually a best practice. Calcium deficiency soon after kidding in the form of milk fever or parturient paresis is not frequent in dairy goats as it is in certain breeds of cows. A reduction of calcium supplementation and replacement of alfalfa hay with grass hay during the dry period can prevent the problem.

Minerals

Several minerals besides calcium require particular attention in proper goat feeding: phosphorus, magnesium, selenium, iron, copper.

Urinary calculi or urolithiasis in male goats are due to nutritional imbalance, especially on high grain feeding with too much phosphorus in relation to calcium and potassium, and more in confinement management than on pasture. Calcium to phosphorus ratio should be 2:1, but grass hay feeding is preferred to alfalfa. Increasing salt in the ration to 4 to 5 percent promotes higher water intake and diuresis. Acidifying the urine with 2 percent addition of ammonium chloride or potassium chloride to the ration also helps.

Selenium deficiency can be suspected in areas with deficient soils, when goats have various reproduction problems, early embryonic death, repeat estrus, retained placenta after kidding, metritis, weak newborn kids. Intramuscular injection with a selenium - vitamin E preparation one month before kidding can prevent symptoms, but addition of 0.2 ppm selenium to the ration provides a more constant protection. Selenium status in goats can be tested best in milk or blood besides hair samples. Iron stores are minimal in newborn kids in contrast to calves. Therefore anemia can be a problem that can be treated with an iron dextran injection or with iron supplementation to the ration.

Magnesium deficiency and grass tetany can occur in early spring grazing on lush pasture, which may be high in potassium, especially cereal grain pastures. Intravenous injection with a calcium - magnesium preparation may be needed to prevent death, but prevention is best by hay feeding prior to turning out to pasture and time-limited grazing of this kind of pasture. A magnesium mineral mix feeding, e.g., 15 percent magnesium oxide in the ration, is also helpful.

Zinc is an element that needs to be supplied continuously, since it is not stored in the body. Blood, milk or hair samples are useless in assessing zinc status of an animal; only rib contents are good indicators. Legumes contain more than grasses, but contents decrease with increasing maturity. Zinc supplementation and treatment helps reduce and cure mastitis, stimulates male reproduction, wound healing, prevents parakeratosis and lameness from foot fissures. Recommended levels are 10-50 mg/kg dry matter daily feed intake. Copper is needed by goats at the level usually provided in dairy cattle or horse rations, in contrast to sheep, which are sensitive to such levels and will develop toxicities. Leaves and certain browse contain more copper than stems of forages, but this will decrease with maturity. Copper deficiencies can be prevented by adding 0.5 percent copper sulphate to the mineral mixture.

Energy

Energy is the nutrient most frequently deficient in goat management, not only of high yielding milkers. Abortions can occur, especially during the time of 90 to 110 days of pregnancy, when undernutrition stresses goats, due to hypoglycemia. Insufficient energy supplies will reduce weight gain and milk yield, but also change the fatty acid composition in the milk fat to less medium chain fatty acids, which are the most desirable fatty acids for human nutrition.

Increasing the energy density of the ration is often necessary as the volume of feed intake is limited, especially in early lactation. Adding fat to the grain ration is increasing energy density effectively as long as it does not interfere with the normal rumen flora. Rumen unavailable or protected fat has been effective at 5 percent supplementation, increasing milk yield, milk fat and protein contents, however, the kind of fat makes a difference in results. Calcium salts of fatty acids are insoluble at normal rumen pH, and reach the abomasum unchanged, where they then can be digested. Another means of increasing energy density of the ration is by pelleting, which improves gains and milk yield by increasing feed intake, but often reduces milk fat content, if effective fiber length is insufficient in the ration.

Protein

Protein is the more expensive nutrient in feeding and therefore often limiting maximum productivity. Industry by-products often are less expensive sources besides the traditional major supplies of oilmeals. However, as forages have higher fiber and lower protein contents with increasing maturity, the least expensive sources of protein are usually forages, alfalfa, clovers, well fertilised grasses, harvested at prebloom or immature stages. Protein supplies to the rumen in the form of degradable protein are necessary for optimum growth of rumen bacteria, but they require energy at the same time, without which some proteins will be wasted into ammonia in the rumen. A minimum of 7 percent crude protein in the diet dry matter is required for normal rumen function, and forage intake will be decreased at lower protein levels. The supply of some rumen protected protein has been effective in increasing milk yield. Excess protein feeding is not only wasting money but is stressing the goat by increasing her blood urea levels, increasing urine excretion and interfering with efficient reproduction. Protein deficiencies will reduce feed intake, rumen function and retard fetal development.

Non-protein Nitrogen

Non-protein nitrogen, such as urea, can be utilised by goats very well, as long as it does not exceed one third of the total nitrogen in the daily diet or 3 percent of the grain ration . A gradual adaptation of at least three weeks is required. Urea may be a cheaper means of providing some of the required nitrogen to goats, but it must not interfere with maximum feed intake. The nitrogen content of feed grade urea is 42-45 percent in contrast to feed protein with 16 percent.

Good urea use in the rumen depends on rations with at least 75 percent TDN and the availability of sufficient starch and sugars like molasses to convert the urea nitrogen into microbial protein, and when the ration protein content is below 12 percent. The addition of alfalfa meal, extra vitamin A and salt helps urea utilisation. Feeding of urease containing feeds like raw beans, legume seeds, wild mustard must be avoided. A common thumb rule is that 6 lb corn plus 1 lb urea equal 7 lb soybean oilmeal nutritionally, but the economics of that relationship have to be calculated to be positive. Aside from grain mixtures, urea is effectively used as a liquid molasses-urea lick or as urea salt block.

Flushing

Flushing is an effective practice of temporarily increased energy and protein supplies in sheep feeding to stimulate estrus in ewes and synchronise pregnancies. This has not been studied much in goats, but practical experience has shown that the principle works in goats as well, making out-of-season estrus, kidding and milk production possible, in addition to increasing litter size.

- — Practical feeding of goats can be grouped into three types:
- — free grazing and no supplementary feeding,
- — limited grazing and supplementary feeding, and
- — confinement feeding with no grazing.

Effects and expectations in meat and milk production obviously differ with these types of feeding under extensive or intensive management. Proper rations must then be calculated differently, depending on the degree of nutrient supply expected from the amount of grazing provided.

GRAZING MANAGEMENT

Under free grazing providing no other sources of nutrients, the grazing strategy must aim towards finding the best pasture in each season without excessive travel and with a stocking rate, that is compatible with good renewal of the vegetation and the best sustainability of forages and browse. The presence of a goatherder will assure this and improve productivity over un-supervised grazing. Nevertheless the nutrient composition varies tremendously from season to season and despite the selectivity of grazing goats, the daily supply often falls short of nutrient requirements of production and at times even of maintenance, so that the goats actually lose milk production, weight and potentially health.

Limited grazing will be a normal consequence by goats when fed supplementary grain. The strategy could be based on the amount of daily milk produced at the rate of 2.5 lb milk per 1 lb grain or more depending on the price of grain to the price of milk ratio.

A superior feeding strategy would be based on body condition scoring. Body condition scoring has been successfully developed for dairy cattle, but applies equally well to dairy goats even in the absence of published suitable picture guides. Body condition score is the visible end result of appropriate or insufficient feeding in relation to production. Out-of-target-range scoring goats will produce less milk and a lower meat price. Reproductive efficiency is significantly reduced by out-of-target-range body condition scores. Also disease frequency is increased.

Confinement feeding

Confinement feeding abrogates any nutrient supply from pasture, although for better health of udder, feet, vitamin D supply from the sun and control of internal parasites some outdoor yards should be provided. The entire nutrient supply must be calculated from composition and requirement tables. Software programs for dairy and beef cattle are available, which have some scaled-down provision on bodyweight. More appropriate would be goat specific programs based on the current NRC and up-dated tables. The University of Wisconsin developed a program, which has not seen widespread use, partly because in the USA no silage is fed to goats as it is the major feed for cows on many farms.

The concept of free choice feeding without rationing to individual goats has been tried successfully . Over a 2-year period 5 Saanen, weighing 133 - 205 lb, produced in 2 lactations from 2,033 - 4,554 lb milk with 3.0 - 3.3 percent fat. Their free choice intake of mixed hay per year ranged from 393 - 459 lb, their grain ration 1,688 - 1,692 lb per year, besides green chop grass, fodder beets and dry beet pulp. The composition of the grain ration was 21 percent crude protein and 10 percent crude fiber. Daily intake between high and low milkers varied from 1 to 8 lb grain; highest daily milk production was 17.8 lb. Production cost analysis in the 2nd year between the highest producer with 4,554 lb milk showed $293.50 for total feed costs vs. $272.19 for the lowest producer with 3,321 lb milk, or $6.44/100 lb milk for the high producer vs. $8.20/100 lb milk for the low producer.

Total mixed ration (TMR) is another approach to free choice feeding, which is very popular in dairy cattle feeding, except that with dairy cattle the major component is silage, mostly corn silage, which is generally not used nor available for goats. Grass silage is fed in Norway routinely and successfully to dairy goats. For many years I have used for my Saanen goats a total mixed ration free choice successfully, and they milked heavy-above 10 lb per day-- even bred out-of-season, kidded twice the year, never had any over-eating disease nor were they vaccinated against enterotoxemia, and had no internal parasite problems despite my not worming them.

A total mixed pelleted ration has been my TMR for years and it is commercially available as a horse maintenance ration, designed for horses, that are neither pregnant nor nursing nor working more than 1 day per week. Thus this ration is supposed to feed horses correctly without letting them get fat. The major composition was 12 percent protein and 26 percent fiber. The high fiber content prevented over-eating by my goats. This pelleted ration was provided to the goats in gravity-flow self-feeders and I have seen it being adopted by the Texas Goat Experiment Station at Prairie View, where turkey big round self-feeders are used for the goats. In addition to this pelleted ration I always provided mixed hay free choice and the goats usually ate less than under conventional feeding, but they preferred stems to get enough fiber. For very high milkers I would feed an extra quarter to half pound of straight soybean oilmeal or sunflower seed at milking time. Individual feeding is the alternative to group feeding and free choice offer of feeds. It is more labour intensive, may save some wasted feed and may better feed according to body condition. It has not been demonstrated whether feeding success in production or profit from the operation is better than in group feeding. Individual feeding requires individual stalls or temporary tie-ups or feeding at milking time or computerized feed dispenser stalls.

Calculating a ration requires seven steps:

1. determine body weight to calculate maintenance requirements of energy, protein, fiber, calcium and phosphorus from tables;
2. determine milk yield and fat content per day plus a challenge factor in early lactation of 10 percent for calculation of production requirements of energy, protein, fiber, calcium, phosphorus from tables;
3. add the two requirement categories for each of the 5 nutrients on a dry-matter basis;
4. determine the composition of your eaten hay (minus the refusals) for the 5 nutrients from tables or actual lab analyses;
5. determine the daily actual hay intake by your goat in question and multiply this with the nutrient composition on a dry-matter basis;
6. subtract the results of step (5) from the total of step (3), giving you the nutrient deficit, which must be provided by a grain supplement on a dry matter basis;
7. determine composition and price of various alternative commercial or farm-grown grain supplements and multiply with the most probable intake level to arrive at the nutrient deficit total, remembering that ration calculations and feeds offered can not exceed the normal level of daily dry matter intake by goats between 3 to 5 percent of body weight. If goats are found to eat less than 3 percent of body weight on a dry-

matter basis, they are either starving or their feed is not palatable to them.

In addition to including the volume capacity of a goat's rumen when calculating rations, one must also consider palatability of the ration and the goat's preference for variety and selection of feeds. Actually voluntary intake is more important than correct nutrient composition. Unless feed intake is maximised, production improvement in the short and long run is not secured.

Feeding Strategies

Feeding strategies under the confinement system can include green chop, agricultural and industrial by-products besides commercial grain rations. This will provide variety, increase intake, lower feed costs, stimulate milk production, but may increase labor costs. Lopping of tree leaves, crop residues from the canning industry like pea and bean vines, fruit pulp, fresh brewers grain, fresh distillers grain, cotton seed, rice, maize, sugarcane by-products, and straw treatment with ammonia or urea have been successfully used in many tropical countries for goat production improvement.

There are many feeding guides now available based on the NRC or similar official foreign tables of requirements and composition. In combination with regular body condition scoring of growing and milking goats, these tables should be adjusted up or down to provide the right supply of nutrients under the circumstances with enough challenge for improved production and growth, or with enough restriction to prevent overconditioning and health risks. If all this is well accomplished then it is time to negotiate the right price for milk, yogurt, cheese and meat from the goats, to proceed with aggressive marketing and promotion to reap the rewards for all this work and to assure that this farm will continue in business for years to come.

56

Diseases of Goats: Prevention and Control

A major concern of goat producers is the diseases affecting their animals. A disease-free herd is more productive and therefore more profitable. However, it cannot be stressed enough that good management practices will practically eliminate all disease processes that could disrupt a flock. Disease etiologies are highly predictable and control methods are well documented. The goat breeder should be able to recognize and control diseases before they become a factor in overall herd management.

Goats do not normally have a high resistance to parasitic infection. Generally, wormy goats will have a rough coat and pale mucous membranes. Open pasture conditions will minimize parasitic infection among the herd members, but parasites can kill. Many of the diseases associated with dairy goats are rare or almost nonexistent in cashmere type goats under any conditions. Those that do affect fiber producing goats and are of concern to managers are caused by bacteria, viruses, protozoa or physiologic dysfunction.

Parasitic diseases area constant source of threat to commercial goat herds. Most animals including humans are prone to parasites however those afflicting goats, such as intestinal nematodes and tapeworms slow growth, ill thrift and possibly prove fatal. Coccidia, intestinal worms and liver fluke have the greatest toll on economic goat-meat production and therefore are central to any goat-care regimen. The key to keeping these under control is good animal husbandry and paddock rotation. Close monitoring and sampling with subsequent testing of fecal egg count is necessary. This will lead to optimizing your drenching program for maximum efficiency.

This chapter discusses some of the more common diseases seen in goats. The discussion is to help identify potential problems goat producers may encounter. It is important to work with your veterinarian and nutritionist to best diagnose and solve nutritional diseases.

Abortions

Nutritional Problems

Proper nutrition is essential for having healthy kids. Simultaneous deficiencies of energy and protein can cause abortion of embryos early in the pregnancy. Deficiencies of some trace minerals such as copper and iodine can be the cause of abortions. Also, excessive selenium for an extended period can cause abortions.

Infectious Problems

An abortion by one or more of the goats in your herd may indicate an infectious disease that needs an overall management response. It is likely that your vet will be required to identify the type of infection causing the problem.

- *Chlamydiosis*: Caused by an intracellular organism. Abortion typically occurs in the last 2 months of pregnancy and especially the last 2 weeks. The rest of the pregnant herd must be considered. Non bred does can catch the infection but it will result in their becoming immune. You should consider injecting pregnant does with tetracyclines by the intramuscular rout to try and prevent them from aborting.
- *Toxoplasmosis:* This is associated with a coccidium of cats. Cats become infected by consuming uncooked meat scraps, placentas, and small rodents. Goats become infected by eating grass, hay or garin contaminated by cat feces. It can result in abortion,

stillbirths and weak kids. However, reducing exposure to cat may help but in may lead to an increase in rats that carry other diseases.

- *Q Fever:* A bacterial disease capable of being transmitted from animals to people caused by *Coxiella burnetii,* a rickettsial organism. *C. burnetii* may be found in sheep, cattle, goats, cats, dogs, some wild animals (including many wild rodents), birds, and ticks. Animals shed the organism in their urine, feces, milk, and especially in their birth products. Abortion or stillbirths occur in late pregnancy, but only when the placenta has been severely damaged. Treatment is with tetracycline. Placentas and aborted fetuses should be destroyed by burning.
- *Brucellosis:* Brucella organisms infect a goats placenta and udder, causing abortion and mastitis When goats in an endemic herd are in a stressful environment and management is not adequate to control nutritional and parasitic diseases, then abortion will occur in the last 2 months of pregnancy.
- *Listeriosis:* Caused by *listeria monocytogenes* a ubiquitous organism that may be found in soil, water, plant litter and digestive tract of ruminants. Abortions occur in the last 2 months. Treatment is usage of tetracyclines.

Caseous Lymphadenitis

Caseous lymphadenitis (CL) is a chronic contagious disease affecting mainly sheep and goats This disease is also called pseudobulerculosis or often "abscesses," and has been referred to as the curse of the goat industry throughout the world.

CL is an infection of goats, caused by *Cory nebacterium* pseudotuberculosis. It is also referred to as "abscesses", because of the peripheral swelling, rupture, and drainage of pus from affected lymph nodes. The prevalence of CL in the commercial goat herds may be as high as 30%. If abscesses affect more than one lymph node, the carcass will be condemned at slaughter. Decreased body weight and milk production also occurs, and reproductive efficiency is often lower when these animals have developed internal abscesses.

Clinical Signs

Most commonly, symptoms are palpable enlargements of one or more of the superficial lymph nodes. The morbidity of the infection rate in goat flocks increase with age, and may approach 70%. The enlarged lymph nodes have a very thick wall and are filled with thick greenish pus. The most common lymph nodes affected are mandibular, prescapular, prefermoral, and supramammary lymph nodes. Less common is involvement of lymph nodes internally in the chest and abdomen. As the animal gets older, abscesses often develop around the lungs, heart, liver, kidney, and spinal cord. They may cause weight loss, pneumonia, and neuoligical signs.

Pathogenesis

C. pseudotuberculosis is spread in the environment by broken and draining external abscesses. The organism survives in the environment for at least one year and can be spread on such items as shearing blades, fences, and feeders. The organism enters the goats body through small breaks in skin or mucous membranes and eventually becomes localised in a regional lymph node. Heavy environmental bacteria contamination occurs in confinement operations and around feeders, goat dairies seem to have a high prevalence of CL. There is some evidence that the organism can penetrate intact skin and mucous membrane.

Diagnosis

Presence of a firm to slightly soft subcutaneous swelling in the location of a lymph node.

Culture: aspiration of the swelling and sending it to the diagnostic lab for isolating and identifying the organism.

Serology: serologic tests such as bacterial agglutination test and synergistic hemolysis—inhibition test are valuable in identifying goats with early stage of the disease (no abscess yet developing). Serologic testing may not be accurate due to the presence of antibodies in previously exposed non-diseased or from cross—reactivity of diagnostic antigens with antibodies against other bacteria.

Treatment

- Separate and isolate the affective animals.
- Ripened abscesses lanced and flushed with 7% iodine solution.
- The pus should be flushed down a drain, or collected and burned.
- Wear gloves to prevent skin infections in humans.
- Wash hands well after handling infected animal.
- Surgical removal of the encapsulated abscess offers the advantages that the treated anials need not be quarantined.
- Antibiotic treatment has not been effective.

Vaccination

A vaccine is available and should be considered in management of CL in infected herds. Vaccine should be considered if you're previously described eradication methods haven't worked or failed. Colorado Serum Company is marketing Caseous-DT, an immunoprophylactic product composed of formalin-killed organisms and toxoided culture supernatant fluid. Which has an efficiency of 70-80% in preventing the clinical manifestations of the disease. The vaccine may cause severe reactions in infected animals, and also interferes with serologic testing for CL.

Coccidiosis

Coccidiosis is a contagious disease of goats, especially young kids, throughout the world. The disease is caused by one or more of approximately 12 different species of protozoa, called Eimeria, which parasitise and destroy cells lining the intestinal tract of the goat. Sheep are also very susceptible to coccidiosis, but even though the sheep forms may share the same names with goat coccidia, many parasitologists believe that the disease cannot be spread from goats to sheep or from sheep to goats.

An infected goat sheds thousands of microscopic coccidial oocysts in its feces every day. When first passed, the oocysts are harmless to another goat. However, under favourable conditions of warmth and mositure, each oocyst matures (sporulates) in 1 to 3 days to form 8 infective sporozoites. If a young kid swallows the sporulated oocyst, the sporozoites are released and rapidly penetrate the intestinal cells. From here on, the life cycle gets very complicated. The coccidia pass through several periods of multiplication during which large schizonts are formed. The intestinal cell of the goat is destroyed and thousands of small forms called merozoites break out and invade other intestinal cells. Eventually sexual stages are reached and new oocysts are produced. The entire life cycle from oocyst to new oocyst takes 2-3 weeks.

If a young kid is suddenly exposed to many sporulated oocysts, it may become severely ill 1-2 weeks later. It will be off feed, listless, and weak. It may show abdominal pain by crying or getting up again as soon as it lies down. At first, the kid might have a fever, but later the body temperature is normal or even below normal. Diarrhea begins pastey, then becomes watery. The kid may dehydrate rapidly. Contrary to various reports written by people more accustomed to calves than kids, the diarrhea is only rarely bloody. Neither is straining common. Signs often show 2-3 weeks after the kids are weaned, because the lactic acid produced by the digestion of milk helps to inhibit occidia in the nursing kid.

Young kids may be killed quickly by a severe attack of coccidiosis. Others—those initially stronger or less heavily infected—will develop a chronic disease characterised by intermittent diarrhea and poor growth. Tails and hocks are dirty. The kid with chronic coccidiosis cannot digest its feed properly because the intestines have been severely damaged. As a consequence, such a kid will be a potbellied poor-doer for months afterwards. Frequently, such a stunted kid will be too small to breed it's first winter.

Even though coccidiosis is typically a disease of the young growing kid, most adults are mildly infected and continuously shed oocysts which serve to infect young kids. Occasionally an adult goat shows temporary diarrhea when stressed or exposed to a new species of coccidia. This is especially common after the doe has been boarded on another farm for breeding.

Diagnosis

Diagnosis of coccidiosis can be based on clinical signs or microscopic faecal exams. Coccidiosis is so common that it should be suspected whenever kids older than about 2 weeks of age are scouring. Sudden dietary changes can also cause diarrhea, but these make the kid more susceptible to coccidiosis. Thus diarrhea that begins with the consumption of too much milk, grain, or lush grass may drag on for days because of coccidiosis. Older kids and adults with diarrhea may have worms rather than coccidiosis, or they may have both problems together. Oocysts can be identified if the feces are mixed with a concentrated sugar solution. The oocysts float to the top, along with larger worm eggs. They are collected and examined with a microscope.

Oocysts may be shed in the feces as early as 10 days after a kid is infected, but often the first attack of diarrhea occurs before oocysts are available to be identified. In these cases, the trained techniciancan do a direct faecal smear to look for smaller merozoites, which do not float in the sugar solution.

If a kid dies of coccidiosis, postmortem examination will quickly give the diagnosis. The small intestine will have many irregular raised white areas, often about 1/8 to 1/4 inch in diameter. A smear taken from these white spots will show many coccidial forms if examined under a microscope.

Whether or not a goat gets sick with coccidiosis depends on several factors. One is the number of oocysts swallowed at one time. Small exposures, frequently repeated, lead to immunity. Large exposures destroy all the intestinal cells at one time and kill the kid. The age of the goat is also important. This is partly because the older animal has usually had time to develop some immunity. Also, very young kids are more fragile creatures. Good nutrition (including vitamin E-selenium supplementation in selenium deficient areas) helps the goat to defend itself against coccidiosis. Immunity to coccidiosis is rarely complete. This means that the healthy adult goat continues to pass many oocysts in her faecal pellets. However, most of her intestinal cells are safe from invading coccidia. As each of the 12 or so coccidia species is completely independent from the others, with no cross immunity, a goat that is happily living with one type of coccidia may develop diarrhea when exposed to a different type.

Prevention

of coccidiosis is very important in larger herds if young kids are to thrive. Once diarrhea has developed, most of the damage to the intestine that leads to stunting has already occurred. Sick kids are treated to save their lives and to limit contamination of the pens, but the owner has already lost control of this contagious disease. Several key facts will help to design a prevention program. The first is that the adult goats are the original source of infection for young kids, because they shed oocysts constantly. All old bedding and manure should be removed from the kidding pens before the new kids are born. Sporulated oocysts are commonly present on the skin of the udder; thus the kid may become infected at the same time as it takes its first drink of colostrum. The doe's udder should be washed and dried before the kid nurses or else the kid should be removed from its dam at once and bottle or pan fed the colostrum.

If only one doe and her kid are present on a farm, and the pens are dry and spacious, coccidiosis is not apt to be a problem. The kids may be safely left with the doe. In larger herds, it is best to raise kids completely separate from the adults until they are ready to breed. Even when rushed from the doe to a clean barn, kids still manage to pick up a few coccidia. As multiplication is rapid, a few can become many very quickly unless good sanitation is stressed. Faecal contamination of feed and water must be prevented. This means that feeders and waterers should be outside the pen whenever possible, and arranged so that faecal pellets can't fall in. Grain should be put in keyhole creep feeders rather than the open troughs that kids love to play and sleep in. Hay racks also must be covered to keep kids out.

Because oocysts have to sporulate to become infective, exposure can be reduced by cleaning the pens daily. Slotted floors are helpful. However, daily cleaning entails a vast amount of work and give disappointing results, if used alone. Ordinary disinfectants don't destroy oocysts. Even to concentrate on keeping the pens very dry, as moisture is necessary for sporulation. Leaking waterers should be fixed at once. Otherwise, the wet ground or floor around the water source is a perfect environment for oocyst sporulation. Small grassy "exercise lots" are also very dangerous and should not be used. It is very important to avoid overcrowding; spreading the kids out decreases the number of oocysts on any given square inch of pen floor or pasture. If many kids are present on the same farm, they should be grouped by age. Putting a 2-week-old innocent kid into a pen with kids 2 months old, where coccidial numbers and immunity have been building up for some time, is to invite disaster for the newcomer. Oocysts are killed by very cold temperatures (far below zero) or by hot dry conditions above 104. Thus, at the end of the kidding season, pens and feeders should be moved out into the hot sunshine for natural sterilization.

Medication

A variety of drugs may be given orally to treat the kid sick with coccidiosis. These include sulfadrugs such as sulfaguanidine and sulfamethazine, tetracyclines (aureomycin or terramycin), and amprolium (Corid R). Each of these has associated dangers if overdosed. Sulfas can cause kidney damage in the kid that is dehydrated. Tetracyclines will interfere with rumen function in older kids and adults.

Very high levels of amprolium may lead to a fatal nervous disease, called polioencephalomalacia, because of a thiamin deficiency. Usually treatment is continued for about 5 days. Labels and veterinary instructions should be followed. If the diagnosis is not certain, and the kid may have bacterial enteritis or pneumonia rather than coccidiosis, sulfamethazine or tetracycline is usually given instead of amprolium.

All of these drugs are coccidiostats, which means that they slow down rather than kill the coccidia. Thus, if a kid is very heavily infected when treatment is begun, medication may not help that kid much. The drugs will greatly reduce the contamination of the environment, and thereby give other kids time to develop immunity. After kids have become immune to the disease they still continue to shed oocysts. Faecal exams may reveal thousands of coccidia per gram of feces. Medicating these older kids or adults will temporarily reduce the passage of oocysts but will not improve growth rate. Within 2 or 3 weeks after medication is stopped, coccidial levels will return to pretreatment values. Thus, except for protection of younger kids, it is a waste of time and money to treat older apparently healthy animals that don't show diarrhea. It is far better to separate the young kids from these older carriers.

Medication of apparently healthy animals is necessary for kids on large farms with previous problems with coccidiosis. The aim is to prevent damage to the intestines rather than waiting for diarrhea to occur. For instance, it may help to treat the kids with anticoccidial drugs on a daily basis for a week or more before stressing them by weaning or moving onto pasture. In some herds, a drug such as amprolium may have to be given daily beginning at 2 weeks of age and continuing until the kids are several months old. Amprolium levels of 25-50 mg/kg daily should be used. This is approximately 10-20 mg per round, and is 21/2-5 times the treatment level recommended for calves. Amprolium is not approved for use in goats in this country. It can be given to each kid individually or it can be mixed with the food or water. As an example, if there are 50 pounds of small kids in a pen, 500 mg of amprolium is mixed with the water, milk or feed that they will consume in one day. The larger kids, by eating more, get more of the drug than do the smaller kids.

Other newer coccidiostats may be mixed with the feed, but most of them have not yet been adequately tested on goats. Rumensin R (Monensin) at 15 ppm in the starter grain has eliminated the coccidiosis problem on at least one large goat farm. This drug is very toxic to horses, so the medicated feed should not be left where a horse can eat it. Another potentially useful coccidiostat, now available only for poultry, is lasalocid. This drug has protected experimental lambs at 2-4 mg/kg/day. The poultry industry has found that the coccidia often become resistant to a drug after 1 or 2 years. Goat owners may also need to change drugs if the one in use ceases to be effective in controlling coccidiosis.

In summary, although most goats carry coccidia and will have positive faecal exams, normally only the young kids become sick with coccidiosis. Deaths and stunted kids result. Raising kids separately from adults, keeping pens clean and dry, preventing faecal contamination of water or feed, and, in some herds, continuous preventative medication are necessary to prevent the disease. It is neither possible nor desirable to completely eradicate coccidia from the adult goats. A low level infection with the parasite serves to keep these goats immune to the disease.

Copper Deficiency

The exact amount of copper required in the goat's diet is currently unknown and is dependent upon several factors. However, the goat needs far more dietary copper than was originally thought. Testing can reveal enough copper in tissue or blood samples and the goat can still be copper deficient. This is due to the complex interaction of minerals in the goat's metabolic system. Copper is essential in the proper development of the central nervous system, correct bone growth, and hair pigmentation. Copper-deficient goats have difficulty conceiving kids and, if bred, abortions are not uncommon. Copper supplementation can sometimes help but cannot always eliminate these health problems.

Copper deficiency can be the result of low levels of the mineral in the soil and in forages raised on the soil; this is primary copper deficiency. However, both the feed and the soil can have adequate copper but its absorption can be interfered with by minerals known as copper antagonists: lead, iron, manganese, various sulphates, cadmium, and molybdenum. This is secondary copper deficiency.

Congenital copper deficiency is the term used to describe the kid who did not receive sufficient copper in utero. Often born swaybacked, the kid stands unsteadily or cannot stand, displays muscle tremors and head shaking, and may grind its teeth. The kid can see, hear, and sometimes can nurse, but he has low blood sugar (hypoglycemia) and sub-normal body temperature (hypothermic). Bone abnormalities are common, particularly in the long (leg and back) bones of the body. Complete recovery from congenital copper deficiency does not often happen because problems that occurred during fetal development may not be correctable. With intensive nursing, swaybacked kids may survive for days or weeks, but they usually do not live long.

Kids who appear to be fine at birth but develop symptoms at around three months of age are said to have the delayed form of copper deficiency. Evidence of atrophied muscles appear, tremors and incoordination occur, and leg weakness is displayed. Usually the problem appears in the rear legs first, but not always. Kids with front-leg weakness will spent lots of time on their front knees. Kids with rear-leg weakness will pull themselves around by their front legs. Death occurs from secondary problems, like pneumonia. This delayed evidence of the results of copper deficiency can be confusing, diagnosis-wise, because it is sometimes accompanied by the neurological form of Caprine Arthritic Encephalitis (CAE), listeriosis, or even muscular dystrophy. Secondary copper deficiency tends to be more responsive to treatment than primary copper deficiency.

Insufficient weight gain, poor appetite, and weight loss are seen in copper-deficient goats of growing age. Adults display more subtle signs of copper deficiency. They are generally unthrifty, anemic poor milk producers, and sometimes have diarrhea. But the most visible sign of copper deficiency in adults is loss of hair colour. Copper is essential for melanin production that causes hair pigmentation. Hair decoloration occurs when copper-containing enzyme is missing.

Other symptoms which may indicate copper deficiency are white or In circles around the eyes; retained placenta; difficulty in conceiving kids (sometimes indicated by continual estrus cycles . . "short cycling" and coming into heat frequently but not becoming pregnant); delayed shedding of hair coat; extreme hair loss; lowered libido in males (lack of crest in breeding); slight hoof deformities; bent legs in yearlings; pink- and other immune-deficiency problems such as frequent bouts with pneumonia, mange or fungus-type lesions, and lice infestation. Copper deficiency may play a role in Floppy Kid Syndrome if the dams were cop- deficient, leaving the kids with only enough stored copper for a week ten days after being born.

Control of proper copper levels in goats is critical. Find out why the goat is copper deficient. Is the soil low in copper? Is there interference in copper absorption because it is binding with other minerals (copper antagonists)? Goats metabolise and store copper much differently from sheep. Do NOT use products labelled "for sheep & goats" because they are woefully insufficient in the amount of copper needed by goats.

Copper can be given to pregnant does and newborns subcutaneously (SQ) in the form of copper glycinate or orally in the does' drinking water via copper sulphate. Severely copper-deficient goats are sometimes given copper boluses which attach to the inside of the body and slowly deliver copper at a predetermined rate.

The easiest and probably the best method, in the opinion of this writer, is to furnish loose minerals with sufficient copper content free-choice to the goats year-around. However, the copper level must be based upon several factors, including the copper available in any ration that is fed to the goats.

It is possible to induce copper toxicity in goats. Copper accumulates in the liver. Red/brown urine may be a sign of copper poisoning. Using calf milk replacers has caused copper poisoning in kid goats.

Up to 1200 ppm of copper may be fed to goats under specific situations. The goat producer must determine what conditions apply in his particular geographic area and based upon his feeding program. Check with a knowledgeable goat veterinarian for proper dietary levels of copper for your goat herd. Much more scientific research needs to be done in this very important area of caprine nutrition.

Diarrhea

Diarrhea is generally a symptom of a digestive system problem. There are numerous known infectious and parasitic causes of diarrhea as well as noninfectious causes. Noninfectious causes can include over-feeding, indigestion, lactic acidosis, copper deficiency and intoxications. The following describes some steps to take for identifying the source of the diarrhea.

— Review any changes to eating or feeding environment. Have you changed the type or amount of feed? Has the animal been grazing in a new or different pasture with different or more abundant food source? Getting off grain to just hay will help.

— Look at the age of animal. Is this a newborn or few weeks old animal? Diarrhea is a common problem during a kid's first few days of life. May be caused by chilling, erratic feeding, dirty bedding, dirty milk bottles and over eating. We use Spectam Scour-Halt (for pig scours) on our young kids with diarrhea and it really does a good job. Diarrhea in a 3-4 week-old kid is likely due to coccidiosis.

— Look at the colour of diarrhea. Green diarrhea can indicate an eating change.

— Check for internal parasites. Worms can cause diarrhea. The feces can be checked for parasite eggs present. This can be done by your veterinary or you can do it yourself. Also review site Veterinary Parasitology. Proper de-worming should take place if parasites are discovered.

Diarrhea should not be considered an illness in and of itself but rather a symptom of other more serious health problems in goats. Before treating a goat for diarrhea, it is essential to determine why the animal is scouring. Diarrhea-controlling medication can make the situation much worse. Slightly soft stool is sometimes the body's way of ridding itself of undesirable products through the purging effect of diarrhea. If the scouring is slightly soft stool, let it run its course.

When body temperature is above the normal range of 101.5 to 103.5 F, use a fever medication and an antibiotic to control infection. Obviously, very watery diarrhea requires a different approach and much more intervention on the producer's part.

There are four major causative agents of diarrhea in goats: bacteria, viruses, parasites, and management practices (i.e., overcrowding, poor sanitation, or nutritionally-induced problems such as over-feeding).

Diarrhea can be the symptom of many different illnesses, including bloat, ruminal acidosis, laminitis/founder, copper deficiency, aflatoxin poisoning, anaphylactic shock, plant toxicity/poisoning, renal failure, selenium toxicity, coccidiosis, enterotoxemia (clostridium perfringens type C&D), salmonellosis, E. Coli infection, caprine herpes virus, heavy parasite infestation, and goat polio. Always run a faecal examination on the goat's feces before attempting any treatment.

However, diarrhea is not always the result of an infectious disease. It can be nutritionally induced by over-feeding on milk or grain, by using poor-quality milk replacers, or by sudden changes in feeding schedules or in the type of feed being offered.

Neonatal Diarrhea Complex, which is the term used to describe diarrhea occurring in kids under one month of age, the cause of which may not ever be diagnosed, usually occurs during kidding season when extremes of weather take place excessive heat or cold or heavy rains. Kids less than one month of age do have not fully funetioning immune systems, so diarrhea can take a heavy toll. Dehydration, acidosis, electrolyte depletion,, and hypoglycemia (low blood sugar) can result. The kid becomes weak and can't stand, has a dry mouth and cold extremities, body temperature drops below normal, and the sucking response is often lost.

Sick kids should be isolated from the herd, placed in sanitary facilities, and fed in containers that are well above ground level to prevent further contamination. Administration of oral and subcutaneous electrolytes along with an appropriate broad-spectrum antibiotic is the recommended treatment.

Coccidia and/or worms usually are the cause of diarrhea in kids over one month of age. Both of these conditions are transmitted by faecal-to-oral contact and occur most frequently in intensive management situations where pens and troughs are not kept clean and dry and where overcrowding exists. Accurate diagnosis of worm or coccidia oocyst infestation is possible only by doing a microscopic faecal count.

Adult-onset diarrhea is less common than in kids, but nevertheless is possible. Overfeeding on grain (such as shell or cracked corn) can cause severe ruminal acidosis... literally shutting down the goat's digestive system... and can result in death. Heavy parasite loads can cause diarrhea in adult goats. Almost anything which negatively affects the proper functioning of the goat's rumen may cause scouring.

Producers should recognise diarrhea as a symptom of a more serious health problem and investigate further to find the cause before running for the medicine bottle. Sometimes, but certainly not always, the diarrhea is helpful in clearing up what is wrong with the goat.

ENTEROTEXEMA

Enterotoxemia is one of the very important diseases and in some areas it is the most prevalent disease of goats. Despite the fact that it is also called "Overeating Disease" it is not caused by overeating. Actually, the cause (etiology) of the disease is the toxin (poison) produced by the bacterium Clostridium perfringens type C or type D.

The bacteria are normally present in the soil and the intestinal tract in relatively small numbers. Under certain conditions the organisms proliferate (reproduce billions of their own kind) in the intestine and produce toxin in lethal quantities. These conditions are those which (1) provide an ideal environment and food for bacterial proliferation and (2) slow down the normal movement of material through the intestinal tract; they are often satisfied by ingesting large amounts of starch when the intestinal tract is not accustomed to it. The disease is often associated with lush fast growing pasture or cereal crops, heavy grain feeding or access to a lot of milk. Illnesses which slow down the intestinal tract, may predispose to the accumulation of dangerous quantities of the toxin.

Most of our knowledge about the disease comes from sheep. There are some important differences in purpose and manner of raising (management systems) that exist between sheep and goats that should be kept in mind when one reads and applies sheep information on goats. First, the disease usually occurs in single (rarely in twins) lambs of a high milk producing ewe; all the milk is consumed by that lamb. In contrast, most goat births are twins, triplets or quadruplets and in the best caprine management systems, the kids are removed from their mother soon after birth.

For a baby kid, treat with 5cc CD antitoxin SQ, 5cc oral penicillin (the long acting) 5cc SQ penicillin, 15cc Pepto Bismol orally. If the kid is in acute pain, you can give .15cc of Banamine and this will also prevent damage to the kidneys and Liver by toxins. Clostridial bacteria growing in the gut are killed topically by the penicillin, which is the drug of choice for the species, Pepto is antacid, anti gas, and reduces pain. Usually give 500mg of thiamin orally also. You must treat the kid for two days, and allow it to nurse if it wants to. On day three, give only the penicillin, and CD antitoxin, and Penicillin SQ, and Calf Pac the kid.

Thus, the type of birth is no factor in goats. Second, dairy kids are seldom fed large amounts of high energy diets for meat production and therefore do not have the same opportunity (unless accidentally or by mismanagement) of exposure to grain. Finally, goats are natural browsers (eat from bushes and trees with their head reaching out or up); they do graze but do not consume as much lush feed (usually in a pasture) as rapidly as sheep. Summarising, the well understood predisposing factors in the sheep disease are not strictly applicable to goats.

The predominant predisposing factors in goats have to do with sudden exposure to grain or large increases in quantity of milk consumed without gradually increasing the amount over several days. This leads to indigestion with slowing of the intestinal tract. This probably plays a large role in the disease in goats by allowing more time for toxin to accumulate within the intestinal tract.

The Type D infection is probably far more common than Type C. The latter type produces a toxin called "Beta Toxin" which causes intestinal necrosis and severe intestinal hemorrhage. It occurs in adult goats.

Epsilon toxin is produced by the Type D bacteria. It produces vascular damage and increases the permeability (openness) facilitating its own absorption. In the animals that die with neither signs nor tissue changes, an extremely large amount of toxin was absorbed very rapidly. When less toxin is produced, the animal lives longer and there is more time for clinical signs and pathological changes to develop.

Signs

In the Peracute disease course, a baby kid may be found dead with no signs or lesions. It may occur after consuming excess feed or after sudden access to highly palatable feed or after prolonged hunger and a normal quantity of feed. The Acute course of disease lasts 4-26 hours and usually ends in death. Initially the temperature may go to 105F with severe abdominal pain (the kid cries so loudly it is best described as screaming).

Profuse slimy or water diarrhea will occur. Depression, wobbly gait, recumbancy (lying down on side often with head down) occur early. Convulsions often occur intermittently and may be accompanied by continuous or intermittent opisthotonos (head thrown straight over back). The animal may slip into a coma before death or die groaning or even crying. These signs occur in kids but can occur in adult milking goats from either Type C or D bacteria.

The subacute disease is more apt to occur in older kids and adults. They may be ill for several days or weeks and show anorexia (refusal to eat) and intermittent severe diarrhea occasionally with epithelial shreds in the feces. They will occasionally eat and with time and appropriate treatment, they will usually recover.

The Chronic form is characterised by intermittent illness lasting several weeks. The goat (usually an adult) will have a dull, stary look, loose feces, an irregular appetite and, if a milker, drop in production.

Tissue Changes

Type C is associated with acute hemorrhagic inflammation and necrosis of the mucosa of the omasum and small intestine. Type D causes mild to moderate (occasionally severe) inflammation and even hemorrhage of the small intestinal mucosa. Petechial hemorrhages may be present anywhere in the body but especially on the epicardium and endocardium. The pericardial sac may contain slight excess of yellow fluid. Microscopic examination of the brain may reveal degeneration of the vascular endothelium with perivascular and intercellular edema with foci of necrosis in several subcortical areas.

Diagnosis

The diagnosis of "enterotoxemia" in goats is probably overdone and is sometimes used to lump any sudden death or acute intestinal disease. The peracute and acute signs are helpful but can also occur with acute salmonellosis or intestinal torsion. Individual or first cases of salmonellosis would probably be diagnosed by post mortem bacteriological examination but if a herd problem exists the history, signs and lesions would justify a presumptive diagnosis. Intestinal torsion is an individual and uncommon event and would rarely be diagnosed ante mortem.

Subacute or chronic cases could resemble coccidiosis, salmonellosis, rumen impaction. Faecal examination, culture and smears would aid in diagnosis of the first two and abdominal palpation, the latter.

The petechial hemorrhages, especially on the epicardium should make one think of enterotoxemia. However, one should look for at least two other signs which together give good presumptive evidence of enterotoxemia; these are, glucosuria and the presence of many short, plump gram positive rods on an intestinal smear.

Ante mortem diagnosis is made early if one can demonstrate a distinct, though transient, improvement of signs after the intravenous injection of 40-100 ml of Type C and D Cl. perfringens antitoxin.

Definitive diagnosis, however, can only be made in the laboratory. Intestinal contents should be preserved by adding 1 ml of chloroform to 10 ml of contents which have been collected in glass within 12 hours of death.

Prevention and Treatment

Vaccination, with Cl perfringens type C and D toxoid by the following schedule along with the good feeding practices of making changes and increases in feed and milk gradually, has provided excellent prevention of the disease. Vaccinate unvaccinated adults twice at 4 to 6 weeks intervals. Vaccinate again during the last month of each pregnancy in order to "booster" her immunity and provide colostral antibodies for the immediate protection of the newborn kids. Vaccinate kids at 2-3 weeks of age and 4-6 weeks later. The older literature suggested that goats produce poor immunity but the results obtained with the alum precipitated vaccine currently in use seem to contradict this idea.

Treatment is ineffective against the peracute and acute cases. However, if ante mortem diagnosis is made one should attempt the use of 50 ml of specific hyperimmune serum intravenously every 4 to 8 hours in the valuable animal. In the subacute and chronic case, antitoxin along with Tetracycline orally at the rate of 5-10 mg/lb (11 to 22 mg/kg) bodyweight will usually effect a cure.

Foot and Mouth Disease

Foot and Mouth Disease (FMD) is caused by a fast-spreading virus, and all cloven-footed animals are susceptible to the disease. Nearly 100% of the animals in an exposed herd will become ill, and young animals may die from the disease. These animals would include cattle, elk, sheep, bison, goats, captive and wild deer, llamas, and domestic and wild swine.

What does, FMD infection look like? Blisters (vesicles) may form in the animal's mouth or muzzle, causing slobbering and drooling. Later, the blisters will break, forming raw patches or ulcers. Blisters and sores also can develop on animal's teats, causing mastitis in dairy cattle. Sheep and goats may have only minimal clinical signs of disease after an incubation period of up to 14 days.

FMD has been diagnosed in 34 countries during the past 18 months. The latest outbreaks have occurred in Great Britain, Northern Ireland, Argentina and France. The continents free of the disease are North America, Australia and Antarctica. Most of the affected countries are still battling FMD. Outbreaks disrupt animal industry, including the export of animals and animal products.

FMD can be transmitted in a variety of way, the most common being direct contact with an infected animal. Once infected, animals become "virus factories," capable of spreading high numbers of viral particles to other animal and into the environment. Infected swine, in particular, can release millions of viral particles when the exhale. While the disease is not considered to be a threat to humans, it's possible for a person to spread the virus to susceptible animals. The disease also can be spread when susceptible animals come into contact with feed, feeding utensils, vehicles, clothing, or holding facilities that have been contaminated with the virus. The FMD virus also can be carried in raw meat, animal products or milk from FMD-exposed or infected animals.

Goat Polio

Goat Polio (Polioencephalomalacia) and Listeriosis (Listeria monocytogenes) are metabolic diseases with similar symptoms that require very different treatments. The goat producer must be alert to the subtle differences in order to know how to treat the sick animal. In most cases, both of these diseases are seen in goats raised under intensive management conditions. Improper feeding, particularly feeding too much grain and too little roughage, is a significant factor in both diseases. Producers pushing the animal to gain weight too fast can induce this often fatal disease in their goats. Sudden changes in feed can also cause the onset of this disease.

Treatment

Polioencephalomalacia is thiamine deficiency. Any change in the rumen's environment that suppresses normal flora activity can lead to decreased thiamine production. Too much grain decreases the pH of the rumen, predisposing the animal to Goat Polio. Thiamine must be present in order for glucose to be metabolised. If thiamine is either not present or exists in an altered form (thiaminase), then brain cells die and severe neurological symptoms appear.

Causes

Causes of thiamine deficiency include feeding moldy hay or grain, overdosing with amprollium (CoRid) when treating for coccodiosis, feeding molasses-based grains (horse & mule feeds), ingesting some species of ferns, sudden changes in diet, the dietary stress of weaning, and reactions to de-wormers Thiabendazole and Levamisole. Each of these can interfere with Vitamin B1 production. Even the usage of antibiotics destroys flora in the rumen and can lead to thiamine deficiency. This is why it is so important to repopulate the gut with live bacteria after using antibiotics or scour medications. Goat Polio is generally seen most often in weanlings and young adults , in contrast to Listeriosis, which most frequently affects adult goats. An increase in Goat Polio occurs in North America during winter, when the availability of forage and quality hay is low and producers start feeding increased amounts of grain.

Symptoms

Symptoms of polioencephalomalacia are excitability, "stargazing," uncoordinated staggering and/or weaving (ataxia), circling, diarrhea, muscle tremors, and apparent blindness. Initial symptoms can look like Entertoxemia (overeating disease). There is a component of "overeating" involved in that the rumen flora has been compromised. As the disease progresses, convulsions and high fever occur, and if untreated, the goat generally dies within 24-72 hours.

Diagnosis

Diagnosis is available via laboratory tests, but the producer does not have the luxury of the time that such tests take. Thiamine is the only effective therapy, and treatment can result in improvement in as little as two hours, if the disease is caught early enough. Thiamine is a veterinary prescription but very inexpensive. Producers should always keep thiamine on hand. Dosage is related to body weight; 10 mg/kg should be given every six hours for at least 24 hours. (One kilogram equals 2.2 pounds.) Initially, IV dosage is best, but SQ or IM can be used. Some producers even give thiamine orally after the initial treatment. If thiamine is unavailable but the producer has multiple B vitamins on hand, make sure the dosage is based upon the amount of thiamine in the multiple B vitamins. The key to overcoming Goat Polio is early diagnosis and treatment. Complete recovery is possible under such circumstances.

Iodine Deficiency Abortion

The Great Lakes basin is deficient in iodine. If goats are not supplemented with iodized salt, iodine deficiency will result. Because the needs of pregnant does are highest (0.8 ppm DM of feed), the most common manifestation of iodine deficiency is abortion or birth of stillborn, and weak kids with enlarged, goitered thyroid glands. The kids may have a reduced hair coat but the most obvious sign are the large bilateral swellings on the neck. The thyroid gland may be several times normal size. Fetuses should be sent to the Animal Health Lab for diagnosis and immediate supplementation of the pregnant does should be done. If iodized salt is already fed, the needs may have increased if the does are grazed on brassica plants (turnips, cabbage, forage rape) while pregnant.

Lactic Acidosis (Grain Overload)

The rumen microflora can only handle gradual changes in forage: grain ratio. If the proportion, absolute amount or type of grain changes too quickly, then lactic acidosis will develop. Feeding order (i.e. grain before forage) also can cause lactic acidosis. The type of rumen bacteria change to gram positive from gram negative and lactic acid is produced. This lowers the pH of the rumen. Once below 5.5, protozoa and bacteria start to die. The acid gets absorbed into the body creating general acidosis. If the pH is low enough, the rumen gets "burned" and, if the goat survives, it often gets secondary rumen and liver infections from bacteria or fungi. Fibre (e.g. hay or silage) is important in the diet as well as it stimulates the goat to chew, thus producing alkaline saliva which serves to buffer the rumen. Diets with little fibre or chopped too finely are more at risk of lactic acidosis.

Simple indigestion may be the first indication of a feeding problem. The goat backs off her feed, usually only for one feeding. If longer than 24 hrs then something else is wrong. Chronic feeding problems will manifest as variable appetite, depressed milk fat and chronic laminitis. Acute laminitis shows up as painful feet. What is more common is the chronic form in which the toes grow abnormally fast with "rings". The quality of the horn is poor and flaky.

Goats may be lame and prone to foot abscesses. Milk fat is depressed because fibre is necessary for the rumen flora to produce the correct volatile fatty acid to make milk fat (acetate). With more severe lactic acidosis, the protozoa die, the rumen becomes static and the goat becomes depressed and dehydrated. The rumen is fluid filled and "sloshy". Diarrhea smells acidic and is yellow in colour.

In very severe cases, there is no diarrhea because of total gut stasis. The goat may appear "drunk" and ataxic. She will go down and will look very similar to milk fever, i.e. cold with dilated pupils. Rumen examination (pH and examination of flora) need to be done to confirm a diagnosis.

Treatment

In severe cases, treatment is heroic and may involve a rumenotomy in which the rumen is surgically emptied out. Supportive therapy includes iv fluids, rumen transfaunation (rumen juice from a healthy animal), alkalinising solutions for the rumen (only done with caution), antibiotics and nursing care.

Prevention

Rations should be formulated and balanced correctly for the correct production group. Forage should be fed before grain and the daily amount divided into at least 3 separate feedings. A total mixed ration (TMR) helps keep the rumen flora happy by not overwhelming them with carbohydrate at any one time. Feed changes all need to be made gradually over several days so the flora have time to adapt. For small holders with a few goats, grain security is an important issue.

Listeriosis

Listeriosis is caused by the bacteria Listeria monocytogenes, which is found in soil, water, plant litter, silage, and even in the goat's digestive tract. The bacteria is known to multiply well in cold temperatures. There are two forms of Listeriosis: one form results in abortions, while the other causes encephalitis. Both types are seldom seen simultaneously in the same herd. The organism can be shed in the milk of both carrier and sick goats. Its zoonotic potential (ability to be transmitted to humans) is of growing concern. Like Goat Polio, Listeriosis is most often seen in intensive management situations.

Unlike Goat Polio, Listeriosis is more common in adult animals than in kids. It is entirely possible to buy infected animals and introduce this disease into a previously uninfected herd, because some goats are carriers who never display any symptoms. Listeriosis is brought on by feeding silage, sudden changes in kind of feed, parasitism, dramatic weather changes, and advanced stages of pregnancy. The encephalitic form is most common, causing inflammation of the nerves in the goat's brain stem.

Symptoms

Symptoms include depression, decreased appetite, fever, leaning or stumbling or moving in one direction only, head pulled to flank with rigid neck (similar to symptoms of tetanus), facial paralysis on one side, slack jaw, and drooling. Diarrhea is presently only in the strain of Listeriosis which causes abortions and pregnancy toxemia. Listeriosis can be mistaken for rabies. Immediate treatment is critical. There is no time to waste with Listeriosis. Recovery is more "iffy" than with Goat Polio.

The exact manner in which both Listeriosis and Goat Polio affect the goat is not well understood at this time. Treatment involves administration of high doses of procaine penicillin every six hours for three to five days, then daily for an additional seven days. Forty-thousand IU per kg of body weight of procaine penicillin is needed to cross the blood brain barrier and put sufficient amounts of the antibiotic into the tissue of the goat's central nervous system. Remember that one kilogram (kg) equals 2.2 pounds.

Prevention

Feed your goats properly. No silage (unless the producer really knows how to use it, and definitely no silage in the hotter and/or wetter climates). No moldy feed or hay. Clean pens. No sudden changes in types of feed. Lots of free-choice quality roughage, particularly in the latter stages of pregnancy.

Ketosis

This is a condition of late pregnancy and early lactation most commonly occurring in the last six weeks of gestation in does with multiple fetuses and in the first 4 weeks in heavily lactating does. Pregnancy toxaemia is a more common condition than ketosis in goats. Through recognition of early signs and symptoms and avoidance of the predisposing factors, it can be reduced to a sporadic condition.

Predisposing Factors

Factors that predispose does to develop pregnancy toxaemia can be divided into two types: inadequate nutrition (they are not offered the correct quantity or quality of the required ration) and adequate nutritional offerings but external or animal factors (e.g. disease) affecting intake.

Inadequate Nutrition

Does carrying multiple fetuses require a much higher level of energy than does carrying singles. These increased needs are compounded by a decreased capacity to consume. When offered ad lib balanced feeds, does in late pregnancy will voluntarily increase energy consumption but expanded uterine contents limit dry matter intake, putting fecund does at even greater risk of developing pregnancy toxaemia. To counteract this, the producer must offer a ration that is more energy and protein dense, e.g. 35% grain to 65% forage. Prolific does on pasture may have difficulty meeting their energy needs through grazing.

Factors Affecting Intake

External factors

Healthy, well fed goats may tolerate many external factors but does chronically underfed are not able to compensate. Inclement weather may interrupt intakes. Feeder space may have been calculated for non-gravid does and may be inadequate for heavily pregnant does. When hand feeding, the producer should be warned to watch for excessive competition. Restricted water intake or poor quality water will reduce dry matter intakes. Rapid feed changes, transport may all temporarily reduce intakes. Poor quality forage that is too woody may also reduce dry matter intake. Preventive treatments such as vaccinating, etc. performed in late gestation may reduce intakes, particularly if they are held off feed for the procedure. Exercise has been shown to increase voluntary intake.

Animal factors

Factors that will reduce intakes include dental disease, old age, smaller body size than group and other concomitant disease such as hypocalcaemia, lameness, gastrointestinal parasitism, Johne's disease, etc. The body condition score of the doe entering into late gestation is important. Does that are very thin (< 2.5) have little fat or muscle reserves to draw upon and are then at increased risk despite a good ration. Very fat does (> 4.0) will readily use body fat reserves in late gestation but also experience decreased voluntary intakes, thus predisposing them to formation of ketone bodies that further suppress appetite.

Clinical Picture

The course of the disease varies but generally develops over three to ten days. A more sudden onset is usually associated with a sudden stress or poor producer observation. Does will start to decrease grain intake, followed by silage and then forage. They separate from the herd, lag behind, and become depressed and gaunt. Other signs of predisposing disease may also be present. Producers that are vigilant when hand feeding does will easily recognise these animals. If missed, the doe may go on to exhibit neurological signs which include an abnormal gait and stance, apparent blindness, stargazing and severe depression followed by recumbency and coma. Some of these signs are attributable to ischemic necrosis of the brain cortex secondary to prolonged hypoglycaemia (low blood sugar). Polioencephalomalacia, hypocalcaemia, toxic mastitis (if near or after kidding), grain overload, listeriosis and lead poisoning are important differential diagnoses. Some producers can smell acetone on the goat's breath (not everybody has this capability). This is the ketone bodies being produced because of the disease.

Clinical Pathology

Most changes are attributable to primary hypoglycaemia resulting from the failure of nutrient intake to meet the combined needs of the doe and fetuses or of the doe's milk production. Circulating glucose demands in late gestation and early lactation are extremely high. The doe must either manufacture this from the production of propionic acid precursor in the rumen or from gluconeogenesis of amino acids derived from the diet or body muscle reserves. This prolonged hypoglycaemia results in suppressed insulin production, which in turn increases fat mobilisation. This is limited by the availability of the glucose precursor oxaloacetate for the tricarboxylic acid cycle. This results in a build-up of acetyl-CoA which is broken down in the liver to the ketone bodies acetoacetate and ß-hydroxybuterate. Low blood glucose is a consistent finding in clinical and subclinical pregnancy toxaemia (often much less than 2.0 mmol/L) but may disappear in cases of severe disease, probably due to the death of the fetuses. Ketonaemia (blood ketones) and ketonuria (urine ketones) are also consistent findings. The serum level of ß-hydroxybuterate (ß-HB) is inversely correlated with serum glucose. Serum ß-HB levels may be used to provide a screening test for flock nutritional status in late gestation. Interpretation of values are presented in table 1 (sheep values). ß-OH levels have been reported at pregnancy toxaemia levels up to 10 days before the ewe develops clinical signs.

Treatment

It is important that the producer recognises the predisposing factors to pregnancy toxaemia and takes action to prevent the disease. Does with reduced appetite and mild depression with no neurological signs, may respond to the following conservative treatment regime: supplementation with propylene glycol (600 mg/ml) at a rate of 60 ml/ BID per OS for a minimum of 3 days; improved nutrition and feeding management; and treatment of any predisposing condition.

Nursing care

Does need to be encouraged to eat and will need extra attention. Correction of other diseases (e.g. if CAE arthritis, a non-steroidal anti-inflammatory could be considered so the doe is willing to get up to eat) This is a disease that needs to be prevented rather than treated. It is generally a management disease and should be initially investigated as a herd level problem rather than an individual sick goat. If one doe is clinically ill, many more in the herd are likely at risk.

Polioencephalomalacia

This is a neurological disease caused by real or relative thiamine deficiency. Thiamine (vitamin B1) is made by the normal bacteria in the rumen. Kids or does on high carbohydrate diets may have an upset in normal rumen flora. A change in bacterial types may cause either a deficiency of thiamine or production of an enzyme which inhibits thiamine activity. The end result is the disease polioencephalomalacia (softening and necrosis of the grey matter of the brain). Overdosing with amprolium (in the treatment of coccidosis), exposure to high levels of sulphur in the diet, or grazing on mare's tail (equisetum) can also result in "polio" but are unusual in comparison to high CHO diets.

Early on in the course of the disease, the goat may show a stiff legged gait. The head may be held high and the animal is anxious. As the disease progresses (often within 6 hours), the goat is blind and the head may be pulled straight back towards its shoulders. The front legs are stiff and the animal may fall down. Once down the abnormal head and neck stance is more evident (opisthotonus). The pupils will constrict to light but the goat will not react to a hand menace. Other rule outs are tetanus (the animal is not blind), pulpy kidney, lead poisoning, listerosis, and other toxins e.g. organophosphates and organochlorines.

Sometimes the only way to make a diagnosis is through a response to treatment. Early polio cases often respond, at least partially if not completely to thiamine administration. Often some response occurs within a few hours of initial treatment. Most other neurological diseases respond slowly or not at all to indicated treatments. Because thiamine deficiency does cause brain necrosis however, time is important. The longer treatment is delayed, the more likely irreversible brain damage may occur. One case may not necessarily mean a herd problem but the feeding management should be reviewed. Some problem herds do require routine thiamine supplementation but first feeding management should be investigated.

Posthitis (Pizzle Rot)

This is a nutritional disease of bucks and wethers that are on high protein diets, often with high inclusion rates of alfalfa hay. The urine is high in urea and, therefore, very alkaline. The urine scalds the prepuce and allow specific bacteria to grow. The prepuce becomes ulcerated, swollen and very painful. The condition is worse in wethers that don't extend their penis to urinate. Affected breeding bucks are often reluctant to extend their penis and may not breed does because of it. Occasionally the condition is so severe that scarring of the prepucial opening mechanically prevents extension of the penis, rendering the buck infertile. Treatment consists of diet change to lower protein rations, clipping the prepucial hairs, cleaning the area and local antibiotics. Pets or valuable breeding bucks may require surgery if scarring is severe.

PREGNANCY TOXAEMIA

This is a condition of late pregnancy and early lactation most commonly occurring in the last six weeks of gestation in does with multiple fetuses and in the first 4 weeks in heavily lactating does. Pregnancy toxaemia is a more common condition than ketosis in goats. Through recognition of early signs and symptoms and avoidance of the predisposing factors, it can be reduced to a sporadic condition. Factors that predispose does to develop pregnancy toxaemia can be divided into two types: inadequate nutrition (they are not offered the correct quantity or quality of the required ration) and adequate nutritional offerings but external or animal factors (e.g. disease) affecting intake.

Does carrying multiple fetuses require a much higher level of energy than does carrying singles. These increased needs are compounded by a decreased capacity to consume. When offered ad lib balanced feeds, does in late pregnancy will voluntarily increase energy consumption but expanded uterine contents limit dry matter intake, putting fecund does at even greater risk of developing pregnancy toxaemia. To counteract this, the producer must offer a ration that is more energy and protein dense, e.g. 35% grain to 65% forage. Prolific does on pasture may have difficulty meeting their energy needs through grazing.

Healthy, well fed goats may tolerate many external factors but does chronically underfed are not able to compensate. Inclement weather may interrupt intakes. Feeder space may have been calculated for non-gravid does and may be inadequate for heavily pregnant does. When hand feeding, the producer should be warned to watch for excessive competition. Restricted water intake or poor quality water will reduce dry matter intakes. Rapid feed changes, transport may all temporarily reduce intakes. Poor quality forage that is too woody may also reduce dry matter intake. Preventive treatments such as vaccinating, etc. performed in late gestation may reduce intakes, particularly if they are held off feed for the procedure. Exercise has been shown to increase voluntary intake.

Factors that will reduce intakes include dental disease, old age, smaller body size than group (see feeder space), and other concomitant disease such as hypocalcaemia, lameness (e.g. CAE arthritis), gastrointestinal parasitism, Johne's disease, etc. The body condition score of the doe entering into late gestation is important. Does that are very thin (< 2.5) have little fat or muscle reserves to draw upon and are then at increased risk despite a good ration. Very fat does (> 4.0) will readily use body fat reserves in late gestation but also experience decreased voluntary intakes, thus predisposing them to formation of ketone bodies that further suppress appetite.

The course of the disease varies but generally develops over three to ten days. A more sudden onset is usually associated with a sudden stress or poor producer observation. Does will start to decrease grain intake, followed by silage and then forage. They separate from the herd, lag behind, and become depressed and gaunt. Other signs of predisposing disease may also be present. Producers that are vigilant when hand feeding does will easily recognise these animals. If missed, the doe may go on to exhibit neurological signs which include an abnormal gait and stance, apparent blindness, stargazing and severe depression followed by recumbency and coma. Some of these signs are attributable to ischemic necrosis of the brain cortex secondary to prolonged hypoglycaemia (low blood sugar). Polioencephalomalacia, hypocalcaemia, toxic mastitis (if near or after kidding), grain overload, listeriosis and lead poisoning are important differential diagnoses. Some producers can smell acetone on the goat's breath (not everybody has this capability). This is the ketone bodies being produced because of the disease.

Clinical Pathology

Most changes are attributable to primary hypoglycaemia resulting from the failure of nutrient intake to meet the combined needs of the doe and fetuses or of the doe's milk production. Circulating glucose demands in late gestation and early lactation are extremely high. The doe must either manufacture this from the production of propionic acid precursor in the rumen or from gluconeogenesis of amino acids derived from the diet or body muscle reserves. This prolonged hypoglycaemia results in suppressed insulin production, which in turn increases fat mobilisation. This is limited by the availability of the glucose precursor oxaloacetate for the tricarboxylic acid cycle. This results in a build-up of acetyl-CoA which is broken down in the liver to the ketone bodies acetoacetate and ß-hydroxybuterate. Low blood glucose is a consistent finding in clinical and subclinical pregnancy toxaemia but may disappear in cases of severe disease, probably due to the death of the fetuses.

Treatment

It is important that the producer recognises the predisposing factors to pregnancy toxaemia and takes action to prevent the disease. Does with reduced appetite and mild depression with no neurological signs, may respond to the following conservative treatment regime: supplementation with propylene glycol (600 mg/ml) at a rate of 60 ml/ BID per OS for a minimum of 3 days; improved nutrition and feeding management; and treatment of any predisposing condition.

Periparturient Hypocalcaemia (Milk Fever)

Hypocalcaemia is usually seen in high producing dairy goats one to three weeks post-kidding and is much rarer than pregnancy toxaemia. Initially the doe is ataxic, nervous and hyperactive but quickly becomes sternally recumbent. The doe stops eating and the ears are cold. The pupils are dilated and respond very slowly or not at all to a flashlight being shone directly at them. The head may be turned back to the flank. Sometimes the hind legs are splayed out behind the doe. The heart is very hard to hear or feel) and beats quickly and weakly. Death follows bloat, regurgitation of rumen contents and aspiration.

The course of disease can be as little as a few hours and occasionally may occur as "sudden death", i.e. the doe is found dead in the morning. Serum calcium levels are decreased, usually less than 1.7 mmol/L (normal 2.1—2.8 mmol/L). To help in diagnosing hypocalcaemia at a postmortem examination, serosanguinous blood obtained from heart clots can be centrifuged and the serum analysed for Ca^{++} levels. The values obtained will accurately reflect pre-mortem values as long as haemolysis and putrefaction have not yet occurred. Again, this disease may look like other diseases and the doe must be examined by a veterinarian in order to differentiate from polioencephalomalacia, advanced grain overload, toxic mastitis, lead poisoning, listeriosis, etc.

Treatment

Clinical cases of hypocalcaemia are usually treated with calcium borogluconate solution (20 mg Ca^{++}/ml) iv and sc. Response should be dramatic. The doe usually starts to shiver and brightens up by the time treatment is finished. If she does not, it may be that the diagnosis is incorrect or is complicated by another disease. It is important that iv treatment only be given in the face of strong clinical evidence of disease.

Prevention

Long term under nutrition is required for primary hypocalcaemia to develop. Goats require calcium rich diets after kidding. Alfalfa hay can provide this. Cereal crop forages such as wheat or oat hay are very low in calcium (0.15% and 0.24% dry matter (DM) basis respectively) as opposed to alfalfa hay (1.4% DM) and should be avoided unless the ration is balanced with other calcium sources. Over-feeding of calcium in late gestation by feeding alfalfa without balancing with anionic salts has been associated with hypocalcaemia in cattle. Feeding an anionic ration in late gestation will also improve calcium absorption from the gut and from the bones. The ration in late gestation and early lactation should also have a calcium: phosphorus ratio of greater than 1.5 to 1. Prevention of pregnancy toxaemia will also help to prevent hypocalcaemia as well.

Sore Mouth in meat goats

Orf is an infectious disease of sheep and goats affecting primarily the lips of young animals. The disease also goes by the names "Contagious Ecthyma, Contagious pustular dermatitis, and Sore-Mouth". The virus may occasionally infect humans and is considered a zoonosis. Orf is caused by a poxvirus. The virus is highly resistant to drying and may remain infectious in dried scabs for up to 12 years. Animals become infected by direct contact with lesions or mucous membranes of infected animals, or by the transfer of virus by contaminated equipment used on the animals.

The disease starts as papules or vesicles (small bumps and blisters) on the skin of the lips, nostrils eyelids, and ears. It may also affect the udder (nursing ewes) or the feet. Eventually the papules and vesicles break open and scab; these are often very sore and may be numerous. As the sores heal the granulation tissue will proliferate and produce masses with a "cauliflower" appearance. The course of the disease is 1-4 weeks with a 2-3 day incubation. Humans become infected with the virus by direct contact with infected sheep or equipment contaminated with body fluids. The disease in humans is usually characterised by a single papular or pustular lesion (small bump/swelling) where the virus enters the skin (e.g. on the hand). The papule usually becomes painful and gradually becomes a firm weeping nodule. Regional lymph nodes may also become swollen. Usually the course of the disease lasts 2-4 weeks following a 3-7 day incubation. More widespread disease may occur as well as severe ocular lesions.

All persons working with infected sheep (the cage or room will be marked) should protect their hands by wearing gloves and washing hands as soon as possible after exposure. All equipment used on sheep should be washed and decontaminated. It should be remembered that gates and pens also may harbour the virus—gloves should be worn at all times working in the animal rooms.

Urinary Calculi

Urinary Calculi—A hard mass of mineral salts in the urinary tract—is a very serious problem normally associated with bucks. In medical books, you will find it under the name of "Urolithiasis". It can result in a painful death in a very short time and is normally the result of some nutritional imbalance. Clinical signs—Painful or difficult urination. Dribbling of urine and animal being vocal. Straining to urinate with possible contractions of the abdominal muscles. Abdominal pain and possibly stretching out and kicking at the abdomen. Grinding of the teeth. There may be crystals found on the hairs around the end of the penis area where the urine normally flows.

Causes

This is most commonly seen in animals that are being maintained in a feedlot environment or pets getting special feed treatment. The makeup of the hard mass of mineral salts varies according to the geographic location of the animals but they are commonly composed of calcium salts and phosphatic complexes.

There are several different types of problems under the common heading of Urinary Calculi::

— *Phosphatic Calculi*—The hard masses are normally formed because of an environment of high-concentrate feed, low-roughage, feed with low calcium-to-phosphorus ratio, high magnesium diets and alkaline urine. A diet mainly made up of grains results in urine with large amounts of phosphorus in it.

— *Silicate Calculi*—This occurs when animals are fed plants grown in sandy areas or water containing high levels of silica.

- *Oxalate Calculi*—Caused by eating excessive quantities oxalate-containing plants such as sugar beet tops and Halogeton, a fast-growing plant of the Western States.
- *Calcium carbonate stones*—Subterranean clover is high in calcium and extensive grazing may cause this.

Clinical Signs

The blocked goat will be uncomfortable and will strain and act depressed. Often the presenting complaint is constipation. If observed carefully, the producer may notice frequent dribbling of small amounts of urine which may be blood tinged. If not sure if the goat is urinating, place in an unbedded, cement pen by himself for several hours. Prepucial hairs may have dried crystals on the end. If not noticed and blockage is total, the bladder ruptures in 24 to 36 hrs. After rupture, the abdomen swells with urine and the goat appears more depressed. He may live another few days before succumbing to the toxins in his system. Occasionally the urethra ruptures and the urine pools under the skin. This condition is called "water belly".

Treatment

If the blockage is at the urethral process, then it can be snipped off. If urine is voided after this "surgery" then the prognosis, while not good, has some hope. Oral therapy with ammonium chloride which dissolves the remaining stones is highly recommended. If the blockage is higher, then there are two options, both with major problems. A perineal urethrostomy, in which the penis is exposed and cut in its location below the rectum and the urethra exposed, often results in failure if the blockage is higher. In addition, even if immediately successful, it is a salvage procedure as the hole heals up in a few weeks and the goat re-blocks. Therefore it is not a suitable option for breeding bucks or pets. The second option is very expensive and few practitioners will undertake the surgery. The abdomen is opened and the bladder opened. A catheter is introduced from the bladder into the urethra and the stones flushed down the penis. If successful, all stones are removed and the buck is still capable of breeding. Often the stones are firmly lodged, there is tissue damage from the stones and recovery is very prolonged. Only valuable bucks or valued pets are recommended for this option.

Prevention

As usual, this is a condition better prevented than treated. The diet should have a calcium: phosphorus ratio of 1.5 to 2:1. Salt should be included at 1% of total dry matter intake. Plenty of fresh, palatable water should always be available. Diets high in potassium should be avoided. Vitamin A requirements should be met. For herds with previous problems, it is sometimes recommended to include ammonium chloride in the ration at ½ % of dry matter intake. This is particularly true with kids on creep grain. Other diseases such as coccidiosis, pneumonia, etc. which might cause decreased water consumption or increased needs may spark an "outbreak" of urolithiasis so these diseases should be managed as well.

White Muscle Disease

White muscle disease is the result of degeneration of skeletal and cardiac muscles of lambs. It is most common in areas where lambs receive good quality feed and are fast growing. It is usually an acute disease that manifests itself in fast growing lambs at three to eight weeks of age. The disease is a manifestation of lack of Selenium, Vitamin E or both. While the leg muscles are generally affected first, any muscle area including cardiac muscle may be affected. Chronic or borderline deficiencies may result in lung edema resulting in increased pneumonia problems. It is quite common for the disease to show up after forced exercise, vaccination procedures, sorting, weaning, fitting or being chased by dogs or children.

Diagnosis is based on clinical symptoms of stiffness, swollen hindquarters or lambs may simply be down. Animals are generally bright and alert and will continue to eat and drink if possible. For those who are mobile, they may have arched backs and appear to have tucked in flanks.

Differential diagnosis from tetanus lambs is made on the basis that tetanus lambs become very rigid and the lambs with white muscle become flaccid. Lambs with intestinal displacement act similar but exhibit pain and generally die within hours, and usually fewer are affected. Laboratory confirmation is achieved by histological examination of affected muscle.

Treatment is accomplished by the use of both Vitamin E and Selenium because the condition may be caused by a deficiency of Selenium, Vitamin E or both. Since the two elements compliment each other both are used in treatment. Selenium is more important in Selenium deficient areas and Vitamin E in Selenium sufficient areas or diets. The injectable Selenium, Vitamin E product that is available does not contain adequate levels of Vitamin E. Once adequate levels of Selenium have been reached, which can be accomplished with one injection, do not repeat because it is very easy to cause acute Selenium toxicity, which will result in sudden death. Additional Vitamin E (200-400 I.U.) should be given with initial Selenium injection. Vitamin E injection should be repeated as needed. Azium (Dexamethasone) helps give symptomatic relief and antibiotics should be used to prevent pneumonia.

Oral Vitamin E products are available that may be added to drinking water. This allows for increased assimilation of Vitamin E and does not cause forced exercise of lambs as injections will. A prevention program is dependent on a definitive diagnosis. Blood and tissue levels of Selenium and Vitamin E are necessary in order to determine the proper supplementation.

Selenium may be provided by injection, feed or salt. We believe that oral administration is safer and more cost effective. In Selenium deficient areas both feed and salt may be utilised. Vitamin E supplementation in fast growing lambs is always essential. It is of particular importance in northern areas with early born lambs. Lambs are particularly susceptible if ewes have been fed one year or older hay.

Bibliography

Amble, V.N., Khandekar, N.C. & Garg, J.N., *Statistical studies on breeding data of Beetal goats*, ICAR Research Series No. 38, Indian Council of Agricultural Research, New Delhi, 1964.

American Dairy Goat Association, Box 865, Spindale, NC 28160, *Dairy Goats-Breeding, Feeding, and Management*, Leaflet No. 439, and Own A Dairy Goat, 1966.

Anon, *Missouri rancher finds 500 goats and 100 cattle fit together well*, sheep! June/July. p. 16, 2001.

Arora, Dinesh ed, *Dictionary of Animal Husbandry*, Biotech Books, New Delhi, 2005.

Barhat, N.K. & Chowdhary, M.S., "Factors affecting some production traits in Rajasthani goats", *Indian J. of Dairy Sci.*, 31: 185-188, 1978.

Belanger, Jerry, *Raising Milk Goats the Modern Way*, North Adams, Mass.: Storey, Carroll, Ricki and Robert, *Cheesemaking Made Easy*, North Adams, Mass.: Storey, 1995.

Bingham, Sam, *A Weed Goat Adventure*, Holistic Management in Practice, November. p. 10-11.

Bubl, Charles, *Meat Goat Production*, Pacific NW Sustainable Agriculture, p. 3-5, 1996.

Chawla, D.S. & Nath, I., "A note on the studies on meat production in goats: effect of castration on dressed meat production in Beetal and its exotic crosses". *Indian J. of Animal Sci.* 49: 406-408, 1979.

Dawra, Sudhir, *Bird Flu: Diagnosis and Treatment*, Biotech Books, New Delhi, 2006.

Devendra, C, *Sustainable Animal Production from Small Farm Systems in Sou*, Daya Publishing House, New Delhi, 1998.

Dunn, Peter, *The Goatkeeper's Veterinary Book*, 3rd ed. Preston, England: Farming Press, 1998.

E B J, *Cows and Poultry in India: Their Care and Management 2nd edn*, Biotech Books, New Delhi, 2004.

Eckles, Clarence H & Ernest L Anthony, *Dairy Cattle and Milk Production*, Biotech Books, New Delhi, 2007.

FAO, *Assessing Quality and Safety of Animal Feeds*, Daya Publishing House, New Delhi, 2007.

FAO, *Good Practices for the Meat Industry*, Daya Publishing House, New Delhi, 2007.

FAO, *Veterinary Education*, Daya Publishing House, New Delhi, 1998.

Farris, Martin, *Producing Goats for the Real World*, Goat Rancher, February 2001.

Gipson, Terry A., "Demand for Goat Meat: Implications for the Future of the Industry", *In: Proceedings of the Fourteenth Annual Goat Field Day*, held April 24, 1999.

Hambidge, Gove ed, *Diseases and Parasites of Sheep and Goats*, Biotech Books, New Delhi, 2004.

Hart, Steve, "Sustainable Brush Control", In: *Proceedings of the Fifteenth Annual Goat Field Day*, Langston, OK. p. 32-35, 2000.

Harwell, Lynn and Frank Pinkerton, Housing, fencing, working facilities and predators, 2000.

Johri, C.B. & Talpatra, S.K., "Growth studies with Jamnapari goats. I: Early growth of Jamnapari goats", *Indian Vet. J.* 48: 389-393, 1971.

Kidwell, Boyd, *Make Money With Goats–No Kidding*, Progressive Farmer. September 2000.

Kumar, Arvind ed, *Animal Science Research: New Dimensions*, Daya Publishing House, New Delhi, 2006.

Kumar, Mahendra, *Animal Byproducts Utilisation through Semi-Moist Rendering*, Daya Publishing House, New Delhi, 2007.

Kyle, Edwin Jackson et al, *Fundamentals of Farming and Farmlife*, Biotech Books, New Delhi, 1999.

Lander, P E, *Feeding of Farm Animals in India*, Biotech Books, New Delhi, 2001.

Lewer, S H, *Wright's Encyclopaedia of Poultry Science*, Biotech Books, New Delhi, 2000.

MacKenzie, Davis, *Goat Husbandry* , 5th Edition, Diamond Farm Book Publishers, Dept. DG, Box 266, Alexandria Bay, NY 13607, 1975.

Malkmus, Bernard, *Internal Diseases of Domestic Animals*, Biotech Books, New Delhi, 2002.

Marwaha, Kavita ed, *Meat Hygiene*, Genetech Books, New Delhi, 2007.

Matthews, John, *Diseases of the Goat*, 2nd ed. Malden, Mass.: Blackwell Science.

Mitchell, Elise, *Brush control with goats*, The Kerr Center for Sustainable Agriculture, Inc., 1991.

Mowat, Noel & Mark Rweyemamu, *Vaccine Manual: The Production and Quality Control of Veteri*, Daya Publishing House, New Delhi, 1999.

Mowat, Noel ed, *Quality Control of Veterinary Vaccines in Developing Countri*, Daya Publishing House, New Delhi, 1998.

Pampori, Zahoor A, *Field Cum Laboratory Procedures in Animal Health Care*, Daya Publishing House, New Delhi, 2003.

Pegler, Henry Stephen Holmes, *Goats and their Profitable Management 9th edn*, Biotech Books, New Delhi, 2005.

Pinkerton, Frank and Bruce Pinkerton, *Supplemental Winter Feeding of Goats*, 1991.

Preston, T R, *Tropical Animal Feeding: A Manual for Research Workers/FAO*, Daya Publishing House, New Delhi, 1998.

Pugh, David A., *Sheep and Goat Medicine*, Philadelphia: W. B. Saunders, 2002.

Sahs, Roger, "Goat Farm Budgeting", *In: Proceedings of the Fourteenth Annual Goat Field Day*, April 24, 1999.

Singh, Vir & Babita Bohra, *Dairy Farming in Mountain Areas*, Daya Publishing House, New Delhi, 2006.

Sinha, S N, *Cow Keeping in India: A Simple and Practical Book on their C*, Biotech Books, New Delhi, 2008.

Smith, Mary and David M. Sherman, "Goat Medicine", Lippincott Williams & Wilkins, Baltimore, MD. p. 538, 1994.

Smith, Mary C., and David M. Sherman, *Goat Medicine*, Malvern, Pa.: Lea & Febiger, 1994.

Speedy, Andrew & Rene Sansoucy eds, *Feeding Dairy Cows in the Tropics/ FAO*, Daya Publishing House, New Delhi, 1998.

Tandon, Veena & Bishnupada Roy, *SEM: Pictorial Guide to Trematodes of Livestock and Poultry*, Regency Publications, New Delhi, 2002.

Triumpho, Richard, *Four-footed bushwhackers*, The Farrow, January. p. 21-22, 2001.

Verma, Rishendra, *Glimpses of Indian Veterinary Science*, Daya Publishing House, New Delhi, 2000.

Index

bl